An introduction to
tropical food science

An introduction to tropical food science

H. G. MULLER

Procter Department of Food Science, University of Leeds

CAMBRIDGE UNIVERSITY PRESS

Cambridge

New York New Rochelle

Melbourne Sydney

Published by the Press Syndicate of the University of Cambridge
The Pitt Building, Trumpington Street, Cambridge CB2 1RP
32 East 57th Street, New York, NY 10022, USA
10 Stamford Road, Oakleigh, Melbourne 3166, Australia

First published 1988

British Library cataloguing in publication data

Muller, H G
An introduction to tropical food science.
1. Food—Tropics 2. Nutrition—Tropics
I. Title
641.1'0913 TX360. T75

Library of Congress cataloging in publication data

Muller, H. G. (Hans Gerd), 1928–
An introduction to tropical food science.

Bibliography
Includes index.
1. Diet—Tropics. 2. Food crops—Tropics.
3. Nutrition—Tropics. I. Title. II. Title: Tropical
food science. [DNLM: 1. Food. 2. Nutrition.
3. Tropical Climate. QU 145 M958i]
TX360.T75M85 1988 641'.0913 87-18410

ISBN 0 521 33488 8 hard covers
ISBN 0 521 33686 4 paperback

Transferred to digital printing 2003

WV

CONTENTS

PREFACE

This introduction to tropical food science is meant for two groups of people. First, there are those living in the tropics who require a simple introductory text. Food science is perhaps the most important science affecting their lives. The second group consists of students, administrators, industrial and research workers in temperate zones who are concerned with food problems but who have no first-hand knowledge of the tropics.

The first six chapters deal with the foundations of food science and are the same all over the world. The next five chapters are a survey of the various foods eaten in the tropics although I have discussed also some of the more important foods eaten elsewhere. In the final chapters I have considered the preparation and preservation of food, its handling in the home and in the food factory. The methods described are basic ones and should be understood by anyone who is involved with food and nutrition.

The sections dealing with health and hygiene generally and water supplies in particular are based on practices used in developed countries and may be difficult to achieve in some tropical countries at present.

I appreciate that insufficient regard for hygiene and the occurrence of nutritional deficiencies are due mainly to lack of knowledge and to poverty. The aim of this book as that of all textbooks is to deal with the former. The problem of poverty must be dealt with through social and economic measures.

I am most grateful to friends and colleagues for advice and photographs, particularly to John Gramshaw, Pat Hayes, Jack Lamb, Helen and Robert Muller, Chris Onuoha, Professor D. S. Robinson, Janice Ryley and Hemant Tewary.

Thanks also to Peter Kunstadt of Atomic Energy of Canada Ltd and Britt Kirstensson of Tetra Pak. I must also not forget Adrian Smith of Leeds University Library, and Lynda Brown who typed the manuscript.

The staff of the Photography Section of the Audio Visual Service of Leeds University prepared several of the pictures for press. I thank them all.

Finally I am indebted to my past and present overseas research students. They provided many of the pictures and taught me much. I dedicate this book to them.

Leeds 1987 H.G.M.

PART ONE

Foundations of food science

1

The tropics: an introduction

Geography and climate

The tropics lie between the Tropic of Cancer and the Tropic of Capricorn (23½°N to 23½°S) but are often extended to include areas as far as 30°N to 30°S where tropical conditions are found. This area which girdles the globe is large and includes Central America, the West Indies and parts of Mexico and South America. Most of Africa lies in the tropics as do much of the Indian Subcontinent, South East Asia and parts of Australia.

The tropical belt is divided into the dry tropics and the wet tropics. The former include the great deserts of North and Southern Africa, Asia and Australia. One often imagines deserts to consist of an enormous expanse of rolling sand dunes but this is not always so. The only characteristic all deserts have in common is that there is very little water during most of the year and indeed it may not rain for several years. Deserts can consist of high mountain ranges with sharp and rocky contours, rolling hills, dried up river beds and, as expected, rippling sand dunes. There are few plants or plant remains (humus) in the soil and when it does rain the water runs off and dry river beds can become raging torrents in little time. It is an odd sight when travelling through the Mohave Desert in the United States in searing heat to come upon signs saying 'Caution Floodwater'. Even at the margin of the deserts plant life is very sparse. One finds eucalyptus and cactus. Of the food plants mainly three are of significance, the date palm, millet and grain sorghum.

From the food scientist's point of view the wet tropics are far more interesting. A typical example is Southern Nigeria, a hot tropical area with high rainfall in the wet season. Among the cereals, maize and rice are found and among the stem and root tubers, yam, cocoyam and cassava. Fruits are very prolific and one finds mango, pawpaw, bread-fruit, coconut, pineapple, peppers, plantain and banana. The main source of oil is the oil palm. Sugar cane is common and in the rubber plantations, particularly in adjacent Ghana, cocoa is found growing in the shade of the rubber plants. A little coffee is also grown.

Climate does not depend only on the distance from the Equator but also on altitude and distance from the sea. Winnipeg in Canada and London in England are much the same distance from the Equator but while London is close to the sea, Winnipeg lies in the centre of a large land mass. In London the climate is never extreme, temperatures rarely exceeding −10 to +30°C. In Winnipeg winters are exceedingly cold, cars leave long condensation trails and if one stands at a bus stop for a few minutes one might feel a tap on one's shoulder and be told 'Excuse me, but your nose is bleeding'. One's nose will then have a white tip and be frozen solid. Winter temperatures of −40°C are not uncommon, but during the summer temperatures rise very sharply to well over +40°C.

The effect of altitude is best shown on Mount Kilimanjaro, the 'White Mountain' of Tanzania. The top at 5000 m is barren and covered with snow and ice glaciers. The zone below, still above the snow line, also grows no plants. There is too little water and oxygen. At 3000 m one finds a typical Alpine desert with mosses, lichens and some grass. Every night there is freezing fog and in boggy moorland areas mosses, grass and sedges are found. Below, at 2500 m the climate is no longer extreme. There are giant heathers 12 m high in a girdle of evergreen trees and tree-like ferns. At 2000 m the climate is that of a hot and humid rain forest. Here live the Chagga people. They find firewood on the upper slopes and collect fodder for their milk cows which they keep in sheds. Potato, maize and bananas are grown as well as beans, cabbages and coffee. At the bottom of the 'White Mountain' in the Savannah live the Masai people. These are herdsmen tending their small cattle, goats and donkeys in the typical meadowland of tropical Tanzania.

It is not just temperature however which governs the climate but also the rainfall. In Nigeria the tropical rain-belt of the south of the country gives way in the north to the rainless desert of the Sahara. From a tropical rain forest one passes to a treeless waste. In the monsoon regions, for instance, in India, the contrast is in the same place. In the dry season there is hard dessicated soil and in the wet season mud and water; as if by magic a desert can turn into a fertile garden. Figure 1.1 shows the tropical zone.

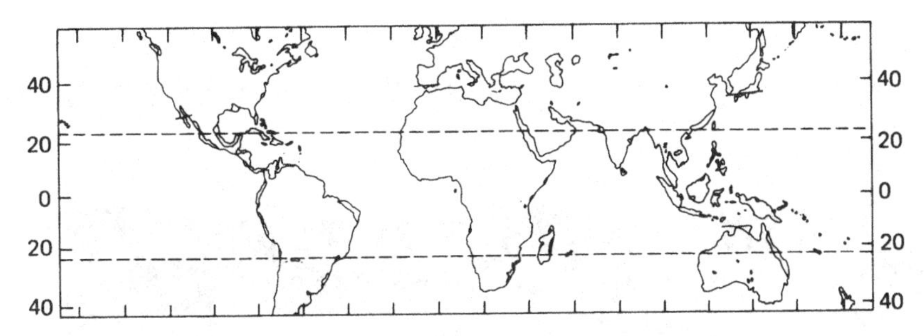

Fig. 1.1. A part map of the world showing the tropical belt.

The effect of climate on food

Storage losses of food in the tropics can be very high. When the weather is hot and dry there is increased wilting of fruits and vegetables. Temperatures in storage bins can reach 50°C. This causes increased respiration of the vegetables with associated weight loss. The heat may also cause increased risk of odour pickup from petrol, paint or tar. This is particularly noticeable with oil or fat-containing foods such as fat meat or groundnuts.

When it is hot and wet, meat, fruit and vegetables may rot very quickly. When relative humidity and temperature are very high, grain and legumes may absorb water from the atmosphere during the relative cool of the night, giving rise to microbial deterioration and ultimately to spontaneous ignition. Drying without special machinery becomes very difficult.

Considerable storage losses can be caused by bacteria, fungi and nematodes as well as insects, birds, rodents and monkeys. Poor road conditions may cause damage to fruit and vegetables by bruising which subsequently leads to rotting. There may be considerable cross infection through domestic waste which may be found in large settements in huge quantities (Fig. 1.5).

Industrial development in the tropics

There is probably a greater range of industrial development in the tropics than anywhere else in the world. To give an example: one can find the most advanced biscuit plant with a computer-controlled band oven as fine as anywhere in the world. There may also be a brick-built tunnel oven fitted with chain conveyors and baking trays similar to the ovens built in the 1930s in Europe. Then one might see a clay oven in which biscuits are

Fig. 1.2. A market scene in Ibadan (Nigeria).

baked at a very low rate of production and under the most unhygienic conditions. A similar range can be found with other food technologies, sales outlets and indeed cities.

One may find a crowded market with an amazing selection of food and little regard for hygiene (Fig. 1.2). The same town may boast modern supermarkets as good as can be found in many western countries (Fig. 1.3). Similarly one can enter a very poor but clean and well kept village (Fig. 1.4), a shanty town (Fig. 1.5) or a modern metropolis such as Singapore or Hong Kong (Fig. 1.6).

Food supply

In the 1960s to 70s the first comprehensive economic appraisal of food in the world was made and one now recognises three areas. The first includes those countries which are either self-sufficient or food-exporting and these present no food problems. Among the developed countries these include the United States of America, Canada and China and among the developing countries, Thailand, Nigeria and Argentina. The second group comprises those countries which require the importation of food but where foreign exchange is readily available through exports. Such countries are for instance the United Kingdom, West Germany and Japan, and, again, there are no food problems. The third group consists

Fig. 1.3. A shopper in a Nigerian supermarket. (Courtesy Unilever plc.)

of those which cannot grow sufficient food and have too little foreign exchange to buy it in sufficient quantity. These are the problem countries.

Usually they are agricultural with a low standard of technology. They tend to be tropical and often have poor soil and climate. They may be subject to natural calamities like drought and plagues of locusts. They are small in size. Although their populations are small (usually below five million) they are increasing rapidly. They have few technological or scientific resources and about half of them have only obtained independence since 1945. Since colonial powers were interested in cash crops, they tended to neglect the food bases of these areas.

In most countries of the world food production has increased in recent decades. However in many developing countries the population increase, largely due to improved medical care, has been so rapid that the production per head has remained steady or has even decreased (Fig.

Fig. 1.4. A typical village in Northern Nigeria

Fig. 1.5. Such slum dwellings with open drains and huge piles of garbage are typical of many towns in developing countries.

Fig. 1.6. Hong Kong Island. An example of a highly developed tropical city. (Courtesy R. E. Muller.)

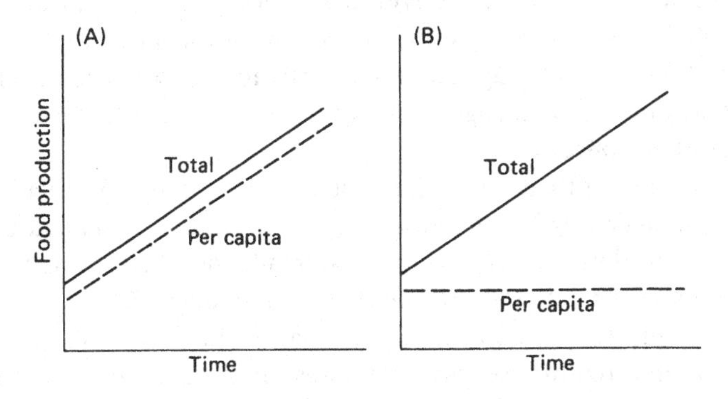

Fig. 1.7. In both (A) and (B) food production has increased equally. In (A) with constant population the per capita food supply has also increased. In (B) due to population increase the per capita food supply has remained steady.

1.7). Acute famines in these problem countries can be relieved by food aid from the developed world but chronic food problems are far more difficult to cure. The problems are various.

Many of these countries have huge debts which must be serviced. Often much of the money borrowed by the developing country promptly flows out again into private accounts in Western banks. Scientific and administrative organisation is often poor. Local senior administrators may be 'well connected' but incompetent. There may be a severe shortage of local trained personnel particularly at intermediate levels. Lack of education and social responsibility may result in apathy and corruption. Particularly in Africa, that sad continent, regional, tribal and religious rivalries hinder progress.

When a country's leaders meet they rarely discuss food problems unless these have given rise to civil unrest. Sometimes there may be a considerable difference in living standards between towns and countryside and the country's elite may aim to achieve western standards and ignore the agricultural basis upon which ultimately the fate of the entire country depends. Instead of growing food for home consumption, resources may be devoted to the cultivation of export crops to finance often unnecessary western imports.

As a rule these export crops fetch less and less money in the international market, while the cost of western manufactured goods increases steadily. This has been called a subsidy paid by the developing countries to the industrialised world. The moneys given in aid are easily swallowed up by the loss in earnings.

International agencies often provide assistance and advisers. Both may occasionally be useless. Frequently at the end of the adviser's contract there is no follow-up although sometimes valuable personal contacts are made. Finally international agencies may not cooperate with each other, to the loss of all concerned.

However, the picture is not entirely bleak. Much food and agricultural research is financed by international agencies in Africa, Asia and Latin America. Considerable efforts are also made on their own by the governments of Brazil, Pakistan and the Philippines. The spread of communications (radio and television) has helped to increase food production and remote sensing satellites have assisted in drought monitoring by assessing the availability of water through the measurement of plant density indicators. Similarly water resource measurement and crop production forecasts can be made using satellites. International financial institutions such as the International Monetary Fund, the Asian and African Development Banks play an ever greater role in the improvement of many poor tropical countries. Last, but by no means least, intermediate technology has been specially tailored for developing countries and some of its aspects are considered in this book.

2

The nutritive value of food

Some definitions

Food is any substance taken by the mouth into the digestive tract which provides energy, material for growth, body repair or reproduction, or regulates these processes.

People in different parts of the world eat very different foods and can still remain fit and healthy. The reason is that almost all foods contain a mixture of materials which are called *nutrients*. Nutrients are the components of the food which are essential. Therefore it does not matter much whether carbohydrates are obtained from yam, cassava, rice, wheatbread, potatoes or any other good source. The same applies to the other nutrients.

A *diet* consists of a food, or more usually a mixture of foods, which is actually eaten, usually in a day. A *balanced diet* contains a sufficient amount of all the nutrients required.

Nutrition is the study of all the body processes which depend on the intake of food while *dietetics* is the study of nutrition in relation to the human body in health and disease.

Malnutrition results when the body receives the wrong amount of nutrients and *undernutrition* when the body receives an insufficient total amount of nutrients. More will be said about these in Chapter 17.

Agriculture deals with the growing of plants for food and with animal husbandry, that is with the rearing of domestic animals. *Food Science* is concerned with the analysis of food and its processing either on a small scale in the home, or on a large scale in the food factory.

Sources of food

Plants take up carbon dioxide from the air through their leaves and water and inorganic salts through their roots. By means of the green pigment chlorophyll they make use of the light energy from the sun to form organic compounds. The reactions involved are very complicated but the overall reaction is simple enough

$$6\,CO_2 + 6\,H_2O \xrightarrow[\text{sunlight}]{\text{chlorophyll}} C_6H_{12}O_6 + 6\,O_6$$

The inorganic elements are also taken from the soil and in this way the whole plant body is built up.

Animals or man cannot use such simple materials for food. They feed either on the plants (herbivores) or on other animals (carnivores). Man can utilise both animals and plants for his food and is referred to as an omnivore.In this way *food chains* are formed. The lowest link in the chain is always a plant and the highest link usually the largest carnivore or man himself. Sometimes the highest link in a food chain is a parasite; examples of such food chains are:

Grass – Goat – Man – Guinea worm
Algae – Shrimps – Whale – Man – Tapeworm.

So through food chains the nutrients are passed along. Sometimes, however, poisons can be passed along as well. For instance, industrial waste containing mercury and dumped in the sea may be taken up by algae eaten by small sea creatures and passed on to fish. Man eats the fish and suffers mercury poisoning as happened not long ago in Japan. Similarly, poisons produced by fungi growing on cereals (e.g. millet) or legumes (e.g. groundnuts) may be eaten by cattle and reach humans through the cows' milk.

Some years ago the testing of atomic weapons caused fallout of radioactive materials. The rain carried these down upon the fields. The grass was contaminated and when cows ate the grass radioactivity was found in their milk. So the testing of nuclear weapons in the atmosphere was stopped by most countries.

Standard measurements and interconversions

The measurements used in this book are for length the metre (m), for weight the kilogram (kg), for energy the Joule (J) and for temperature the degree Celsius (°C).

Table 2.1 *The weights of various foods (g) measured in the same milk tin (180 ml)*

Cornflour	113	Millet	161
Gari	119	Sorghum	146
Salt	250	Maize	167
Sugar	175	Rice	170
Black eye beans	140		

Other units used in this book are derived from these.

1 kg = 1000 gram (g), 1 g = 1000 milligram (mg)
1 mg = 1000 microgram (μg)
1 m = 100 centimetres (cm), 1 cm = 10 millimetres (mm)
1 mm = 1000 micrometres (μm)

Since the Joule is too small a unit for the nutritionist to use the kilojoule (kJ) equal to 1000 joules is used; 1000 kJ are equal to one megajoule (MJ).

Volumes are given in litres (l). These are subdivided into millilitres (ml) where 1 l = 1000 ml (see also Appendix 2).

This system of measurement is used in scientific work, and indeed for trade in most parts of the world. However sometimes the old British system of pounds and ounces is still employed and in Central America the American units of cups, pints and quarts. Common measures in West Africa are the gin bottle (750 ml), the squash bottle (750 or 1000 ml) and the milk tin (180 ml).

When measuring solids by volume much depends on the density of the solid. For example a heaped milk tin of gari (see later) is much lighter than one containing salt. Water (moisture) content also affects the density of food.

Buying solid food by volume is always inaccurate. Table 2.1 gives the weights of various foods in a heaped milk tin. Sometimes the milk tin used is of a different size and that makes matters worse.

The weight of wrapping is also important. Table 2.2 shows the weight of samples of kenkey (see later) purchased in various markets in Ghana. The Table shows that both the edible weight of kenkey and the weight of the leaf wrapping varied a great deal. The wrapping leaves from maize or plantain must be deducted from the edible matter.

The nutrient value of food must now be considered.

Table 2.2 *Weight of kenkey with its leaf wrapping purchased in various towns in Ghana (g)*

Type of kenkey	Town of origin	Total weight (g)	Wrapping (g)	Edible weight (g)
Kokui	Kpandu	270	15	255
Mbor	Winneba	440	10	430
Dokon pa	Winneba	575	25	550
Dokon pa	Yanoransa	525	35	490
Ashanti	Osino	330	50	280
Nsihu	Cape Coast	440	10	430
Ntaw	Amosiama	550	10	440
Ga	Accra	320	40	280

Water

Water is the most important nutrient of all. If water is withheld for only a few days a person dies. About 60% of body weight is water. Most of that is within the cells of the body (40%) and the remainder (20%) is found in blood, lymph, spinal and synovial fluids and the various secretions such as sweat, tears and saliva.

Water is essential as a lubricant and solvent in digestion and assimilation (see Chapter 3) and important for temperature regulation. Extensive sweating cools down the body by evaporation.

Water requirements vary widely and as a rough guide an adult requires about 1 l per 4000 kJ of food intake. Therefore an adult requires about 2–3 litres of water per day. Young people may require up to 50% more because they tend to have less fat in their bodies. Their body water content is higher since fat contains no water.

Not all water is obtained by drinking liquids. At least 25–50% of water is obtained through solid foods. Only very few foods contain no water at all. Table 2.3. shows the water content of some tropical foods.

A regular supply of water is required to make up for water loss from the body in the form of faeces, urine, sweat and breath. Normally the kidneys regulate the water but extra losses may occur through vomiting, diarrhoea or heavy sweating.

The water content in a food is determined by weighing a sample before and after drying to constant weight. Either an air oven is used at 135°C or a vacuum oven at 95–100°C. The test usually takes 1 h in the air oven and 5 h in the vacuum oven.

Fig. 2.1. Water is man's most important need.

Fig. 2.1a. Drinking water for sale in Egypt.

Table 2.3 *Water content of some common tropical foods (g/100 g edible material*

Beer, fruit juices	85–95
Fruits (e.g. mango, pawpaw, pineapple)	80–90
Cereal porridge	50–80
Fish, fresh	80
Roots and tubers (e.g. yam, cassava)	60–75
Meat, fresh lean	70
Rice, boiled	70
Kenkey	65
Wheat bread	40
Meat, very fat	25
Fish, dried	20
Cereals, dry (e.g. rice, sorghum)	10–15
Milk powder	5
Biscuits, chocolate	2–5
Sucrose, cooking oil	0

The sources of water and the diseases associated with it are considered in Chapter 17.

Proteins

Proteins are very large molecules consisting of long strings of *amino acids*. These strings may be coiled up into balls or lie side by side to form flat sheets.The human body consists largely of proteins and these cannot be made up in the body except from amino acids taken in the food. Since proteins are part of every living cell almost all foods with the exception of sugar contain protein. Good animal sources of protein are meat, fish, eggs and dairy products. Good plant sources are peas, beans and groundnuts.

There is not just one substance called 'protein' but many thousands, all slightly different. When heated the protein of egg coagulates but the chief protein in milk will remain in solution. Some proteins dissolve in water, some in dilute salt solutions, some in dilute acids or alkalis and some in none of these solvents.

The reason for this different behaviour is due to the fact that (1) proteins are composed of different amino acids, and (2) the amino acids are bound in different ways.

All amino acids contain the elements of carbon, hydrogen, oxygen and nitrogen, and the amino acids methionine and cysteine also contain sulphur.

There are about twenty commonly occurring amino acids, all of different composition. Not all amino acids are present in any one protein: zein, the chief protein of maize, contains no lysine, gelatin from meat no tryptophan and insulin no methionine.

Some amino acids which the body requires can be produced from other amino acids by interconversion. Other amino acids, however, cannot be made in the body and must be obtained ready made. These are called essential amino acids. Table 2.4 shows the more important amino acids. The essential ones are marked with an asterisk. Histidine is essential to infants only. Adults can manufacture histidine from other amino acids.

The most important bond connecting the various amino acids in the protein chain is the *peptide bond*. All of the amino acids possess at least one amino group and one carboxyl group. The peptide bond (–CONH–) is formed by a reaction between an amino group of one amino acid and a carboxyl group of another.

$$\begin{array}{c} COOH \\ | \\ R-CH \\ | \\ NH_2 \\ + \\ COOH \\ | \\ R'-CH \\ | \\ NH_2 \end{array} \longleftrightarrow \begin{array}{c} COOH \\ | \\ R-CH \\ | \\ NH \\ | \\ C=O \\ | \\ R'-CH \\ | \\ NH_2 \end{array} + H_2O$$

In this way long chains are built up with the amino acids in a definite, genetically determined order. This sequence of the amino acids in the chain is referred to as the primary protein structure. This in turn gives rise to a localised order of small sections of the chain which is stabilised by intra-chain bonds (i.e. cross-links *within* the chain). In this way a helix, a sheet, or other configurations can be produced. Further interactions between the secondary structure give rise to a very compact configuration, the tertiary structure. Finally, one compact structure can form inter-chain links (i.e. cross-links *between* different molecules) with other molecules and so give rise to a quarternary structure.

Figure 2.2 shows the molecule of lysozyme found in egg whites. There are 129 amino acids of 20 different kinds. The letters identify the amino acids. Sometimes cross-links are found. In this instance the cross-link is formed by cystine.

Table 2.4 *The structure of the important amino acids*

Group	Amino acid	Structure
Monoamino monocarboxylic	Glycine	$H{-}CH(NH_2){-}COOH$
	Alanine	$CH_3CH(NH_2){-}COOH$
	* Valine	$(CH_3)_2{>}CH{-}CH(NH_2){-}COOH$
	* Leucine	$(CH_3)_2{>}CH{-}CH_2{-}CH(NH_2){-}COOH$
	* Isoleucine	$CH_3{-}CH_2{-}CH(CH_3){-}CH(NH_2){-}COOH$
Monoamino dicarboxylic (acidic)	Glutamic acid	$HOOC{-}CH_2{-}CH_2{-}CH(NH_2){-}COOH$
	Aspartic acid	$HOOC{-}CH_2{-}CH(NH_2){-}COOH$
Diamino monocarboxylic (basic)	Arginine	$H_2N{-}C({=}NH){-}NH{-}CH_2{-}CH_2{-}CH_2{-}CH(NH_2){-}COOH$
	* Lysine	$H_2N{-}CH_2{-}CH_2{-}CH_2{-}CH_2{-}CH(NH_2){-}COOH$
Heterocyclic	Proline	$H_2C{-}CH_2$ / H_2C / $CH{-}COOH$, ring closed through $N{-}H$
	Hydroxyproline	$HO{-}C(H){-}CH_2$ / C / $CH{-}COOH$, ring closed through $N{-}H$
	* Histidine	imidazole ring (N, NH) $-CH_2{-}CH(NH_2){-}COOH$
	* Tryptophan	indole ring (N–H) $-CH_2{-}CH(NH_2){-}COOH$
Aromatic	Tyrosine	$HO{-}C_6H_4{-}CH_2{-}CH(NH_2){-}COOH$
	* Phenylalanine	$C_6H_5{-}CH_2{-}CH(NH_2){-}COOH$
Hydroxyl-containing	Serine	$HO{-}CH_2{-}CH(NH_2){-}COOH$
	* Threonine	$CH_3{-}CH(OH){-}CH(NH_2){-}COOH$
Sulphur-containing	Cystine (and cysteine)	$S{-}CH_2{-}CH(NH_2){-}COOH$ / $S{-}CH_2{-}CH(NH_2){-}COOH$ (S–S linked); $HS{-}CH_2{-}CH(NH_2){-}COOH$
	* Methionine	$CH_3{-}S{-}CH_2{-}CH_2{-}CH(NH_2){-}COOH$

Determination of proteins

The crude protein content of food is determined by measuring the nitrogen content of the amino acids. The best known of these methods is the Kjeldahl method. Here the dried sample is digested by boiling with

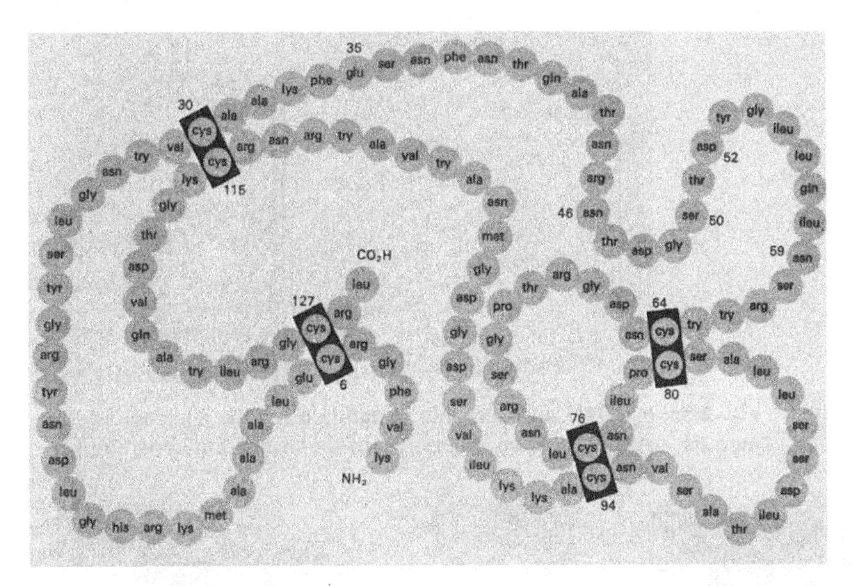

Fig. 2.2. A chain of amino acids making up the giant molecule of lysozyme. (Courtesy Professor Sir David Phillips.)

concentrated sulphuric acid for two to three hours. The temperature of the boiling mixture should be 370°C and this is achieved by adding 0.8 g of potassium sulphate per 1 ml of sulphuric acid. Usually 1 ml of acid is sufficient for 1.5 mg of nitrogen in the sample. A catalyst such as copper sulphate is added to promote the reaction. During the digestion the nitrogen of the amino acids is converted into ammonium sulphate. The digest is then cooled and diluted with water. The digestion arrangement is shown in Fig. 2.3a.

The digest is now made strongly alkaline by the addition of 40% sodium hydroxide solution and the ammonia produced is distilled over and collected in sulphuric or boric acid (Fig. 2.3b, c). The acid is then titrated and 1 l of 1 M HCl is equivalent to 17 g ammonia or 14 g nitrogen from the original amino acid mixture in the protein.

If the proportion of nitrogen in a particular food is known then multiplying the nitrogen value by a suitable factor will give the protein content. In most food proteins there is 16% nitrogen and therefore the factor to change the nitrogen content into protein content is 100/16 or 6.25. Not all food proteins contain 16% nitrogen and therefore some foods have different factors. Examples of these are given in Table 2.5.

The Kjeldahl method is not completely accurate because not only the amino nitrogen is measured but other nitrogen as well. The latter may be contained in amines, nitrates, nitrogen-containing sugars (N-glycosides)

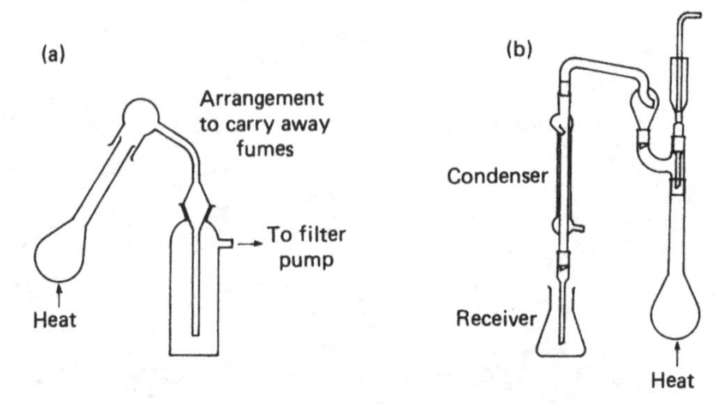

Fig. 2.3a, b. Total nitrogen determination using the Kjeldahl method: (a) the digestion assembly, (b) the distillation apparatus.

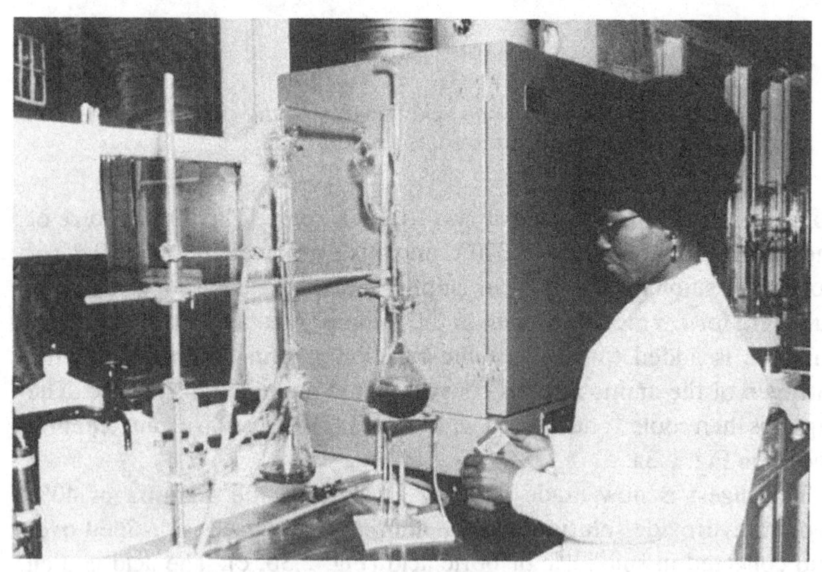

Fig. 2.3c. The Kjeldahl distillation apparatus in use.

and fats (glycolipids) as well as B-vitamins. Nevertheless the error is small. Other methods of protein determination are available and some depend on colour reactions. An important one is the Biuret reaction where cupric ions at alkaline pH react with the peptide bond. There are also a number of reactions with sulphur-containing dyes. Some of these reactions are the basis of commercial instruments.

There is also the formol titration. Here formalin added to a neutralised aqueous protein solution reacts with the amino group with the release of a proton which can be titrated.

Table 2.5 *Protein factors for various foodstuffs. (Multiply the Kjeldahl nitrogen by the factor given to obtain crude protein.)*

Foodstuff	Conversion factor	Foodstuff	Conversion factor
Milk and milk products	6.38	Almonds	5.18
Wheat	5.83	Other nuts	5.30
Rice	5.95	Soya	5.71
Groundnuts, Brazil nuts	5.41	General factor for other foods	6.25

Table 2.6 *Protein efficiency ratios of various proteins*

Algae	1.90	Some other legumes	1.1–1.7
Yeast		Root crops	0.7
(*Saccharomyces*)	2.2	Shrimps and prawns	2.2
(*Candida*)	1.0	Meat and Fish	2.3
Soya Beans	2.3	Egg	3.9
Groundnuts	1.7	Cow's milk	3.1

However all these reactions are standardised against the Kjeldahl test.

The most recent method is based on near infrared radiation but the equipment used for routine analysis is exceedingly expensive.

While these methods give an indication of the total amount of protein present in food, they tell us nothing of protein quality. Obviously a protein containing all essential amino acids is of more use to the body than a protein which lacks some or indeed most of them. Routine chemical analysis of all the amino acids in a protein is now possible but the best methods are biological ones. There are quite a number of these but the simplest is the determination of the protein efficiency ratio (PER).

Because the proportions of essential amino acids required by man are similar to those required by the rat, that animal is usually used for the test. It is fed on a diet containing sufficient energy, vitamins and minerals but the only protein source is the one under test. During the next two weeks weight gain and protein intake are measured. The PER is defined as gain in weight divided by protein intake. Table 2.6 gives some values of PER. The lowest value is 0 for a very poor protein, the highest just over 4.

Some important reactions of proteins

Denaturation. Two reactions of proteins are nutritionally important. The first is denaturation. Most amino acids and proteins are soluble in water. When they are heated, subjected to ultraviolet light or intensive mixing their structure may be altered. This results in a change of properties. An example is adding acid to a protein in solution which as a result may become insoluble and precipitate. Whipping egg white into a foam or boiling an egg are other examples of protein denaturation.

Hydrolysis. The other reaction of nutritional importance is hydrolysis. The bonds between the amino acids making up the protein can be broken by acids or alkalis. In the human digestive system hydrolysis is carried out by enzymes (Chapter 3). During hydrolysis the amino acid chains are broken down into smaller and smaller units. Eventually the free amino acids are able to pass in solution through the wall of the alimentary canal into the body.

The nutritional value of proteins is not normally affected by heat, acids, alkalis or light. Protein content may be increased through microorganisms produced during fermentation (Chapter 14). Protein uptake may be decreased by tannins (Chapter 7).

Enzymes

Enzymes are proteins which catalyse chemical reactions in living cells. They were first demonstrated in yeast and the term 'en-zyme' means 'in yeast'.

Because enzymes have a unique protein structure they are usually quite specific: they catalyse one type of reaction only. Some enzymes depend entirely on their structure for their effect but others require additional cofactors which can be metal ions, complex organic chemicals and sometimes vitamins.

Enzymes do not make reactions possible which are chemically not feasible. They simply reduce the energy required to make the reaction happen. For instance to break down hydrogen peroxide (H_20_2) to hydrogen and oxygen requires 18 Cal/mol without the enzyme, but only 2 Cal/mol in the presence of the enzyme catalase.

Enzymes are affected by both heat and pH. They usually work best at the pH of the medium in which they are normally found. Intracellular enzymes usually work best at about pH 7 while gastric enzymes which are found in the acid stomach content work best in an acid medium (Fig. 2.4a).

As with all chemical reactions their reaction rate increases with increasing temperature. However, soon the protein part begins to become denatured and the enzyme action decreases rapidly. The

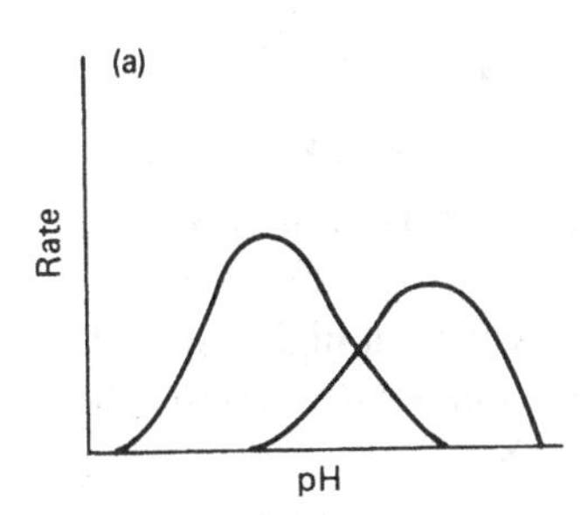

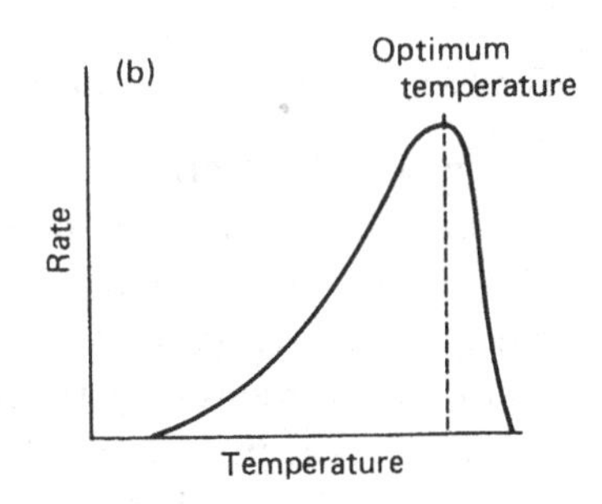

Fig. 2.4. Effect of (a), pH and (b), temperature on enzyme rate.

Table 2.7 *The six major classes of enzymes*

Name	Reaction catalysed	Equation
Hydrolases	Hydrolysis	$AB + H_2O \rightarrow AOH + BH$
Transferases	Transfer of a group	$AB + C \rightarrow A + BC$
Oxidoreductases	Oxidation–reduction	$ABH_2 \rightleftarrows AB + 2H$
Isomerases	Intramolecular rearrangement	$ABC \rightarrow ACB$
Lyases	Remove groups or add to double bond	$ABC \rightarrow AB + C$
Ligases	Join molecules	$A + B \rightarrow AB$

temperature at greatest rate is referred to as the optimum temperature (Fig. 2.4b). It is usually in the region of 50–70°C but occasionally may be as low as −15°C or as high as +80°C. Some of these pecular enzymes can be very useful technologically (Chapter 16).

Some enzymes can be inhibited by heavy metals such as mercury, lead or selenium or by hydrogen cyanide. That is why these substances are such powerful poisons.

Some enzymes also require activation before they can act. An example is the activation of pepsinogen in the stomach, which is described in Chapter 3. The use of enzymes in food is considered in Chapter 16.

Enzyme names are given the ending '-ase'. There are six major groups which are shown in Table 2.7.

Enzymes isolated from their cells can also be immobilised on water-insoluble materials such as silicious materials or organic polymers and still retain their activity. They can then be packed into columns and as a solution is percolated through, its chemical composition is changed. In this way a protein solution can be changed into one containing peptides and amino acids. Other examples are given in Chapter 16.

Enzyme-linked immunosorbant assay (ELISA)

This is a very new and powerful analytical method which is based on the immune response in the blood of animals. When foreign materials e.g. bacteria, viruses or alien proteins get into the blood, the body tries to eliminate them. The foreign body, the antigen, causes a highly specific reaction giving rise to an antibody designed to eliminate the invasion. The ELISA test involves an enzyme labelling of the antibody and its binding (immobilisation) on special plastic plates or columns. The linked enzyme causes a colour reaction which is measured. The preparation of the reagents requires high technology but that is built into the molecule and not into the equipment of the user. The test kits are relatively cheap and can be used to distinguish meats (e.g. pork from beef), indicate toxins (e.g. aflatoxin) or demonstrate dangerous microorganisms such as *Salmonella* or *Staphylococcus* (Chapter 6). (See Appendix 3.)

Carbohydrates

The name carbohydrate means 'carbon hydrate' and it comes from the fact that the carbohydrates consist of carbon, hydrogen and oxygen. The latter two are in the same ratio as in water. So the general formula for carbohydrate is $C_x(H_20)_y$.

Just as the proteins are built up of long chains of amino acids, so the higher carbohydrates starch and cellulose are built up of chains of sugars. The most important sugars contain six carbon atoms, and are called hexose sugars (hexa means 'six' in Greek). Glucose and fructose are examples. These two sugars are monosaccharides ('mono' means 'one', 'sacharon' means 'sugar' in Greek).

Sucrose consists of two molecules, one of glucose and one of fructose. Therefore sucrose is a disaccharide ('di' means 'two' in Greek). These components can be split apart by acid hydrolysis or by the enzyme invertase. This process is called 'inversion' (p. 163).

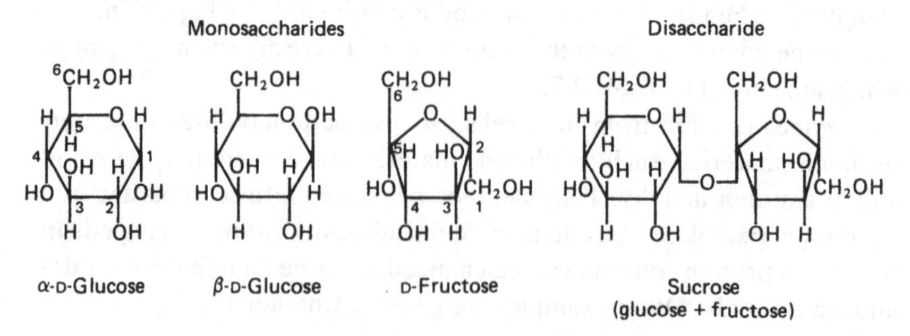

Glucose, fructose and sucrose are soluble in water and easily absorbed through the intestinal wall.

Glucose units can be built up into long chains in two ways to form either cellulose or starch. Both are very large molecules but because they are bonded in different ways they have different physical and chemical properties. Both are called polysaccharides ('poly' means 'many' in Greek).

Polysaccharides

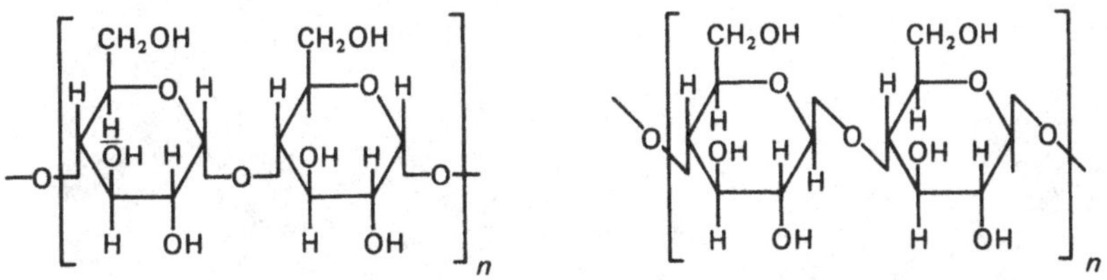

Starch

Cellulose

Cellulose is a straight-chain polymer of 1–4, β–D-glucose. It forms the major part of the fibrous material of which plant cell walls are made (Fig. 2.5) and is not found in animals. Cellulose is insoluble in water and very resistant to digestive enzymes. Together with woody material (lignin) it is referred to as crude fibre. This is determined as follows.

First the fat or oil is extracted from the food with ether (see later) and the residue is then boiled for 30 minutes with 1.25% sulphuric acid. The solution is then filtered and the residue washed with water, then boiled for 30 min with 1.25% sodium hydroxide, rinsed with water and dried. The dried residue is weighed and burnt at 600°C for 30 min. The mineral matter remaining is known as ash. The weight of dried residue before ashing less the ash itself is referred to as crude fibre.

Crude fibre is very similar to that portion of the food which is not digested and absorbed by the body. A more accurate measure is dietary fibre. This is determined in a more complicated way involving extraction with methyl alcohol, ether, dilute and concentrated sulphuric acid as well as digestion with the enzyme diastase.

Starch forms the major carbohydrate reserve of plants. Seeds and underground storage tubers and roots are the most important sources for man. It consists of varying proportions of amylose and amylopectin. Amylose is a straight-chain polymer of at least 200 α–D-glucose units bonded in the 1–4 position. Amylopectin is a branched polymer of α–D-glucose bonded in both the 1–4 and the 1–6 positions. The latter bonding causes the branching, as shown below.

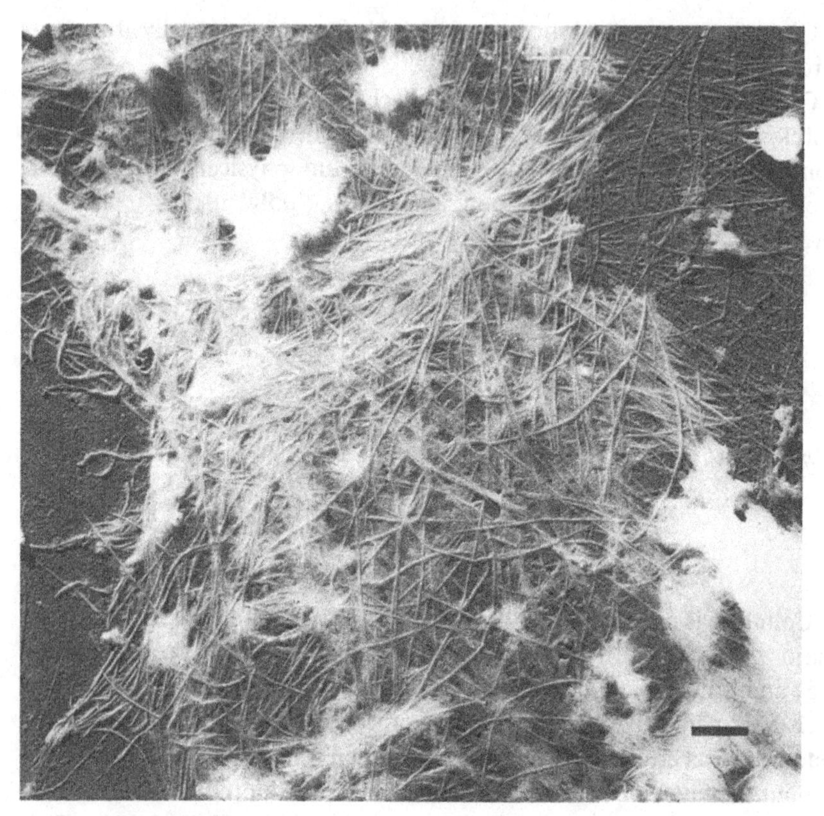

Fig. 2.5. Cellulose fibrils in a primary plant cell wall; bar = 1 μm. (Courtesy B.J. Nelmes.)

1-6 bond

Glycogen, the carbohydrate storage product of animals, is even more closely branched than amylopectin. Table 2.7a gives the percentage of amylose in various starches.

The starch is arranged in granules which very often have very distinctive shapes which can be identified under the microscope (Figs. 2.6a–d). In order to release the starch for digestion in the body the starch grain is

Table 2.7a *Percentage of amylose in starch from various sources*

Origin	Amylose (%)
Sugary mutant maize	70
'Steadfast' pea	67
Chick pea	33
Barley	27
Wheat	25
Cassava	18
Waxy barley	3
Waxy maize	0–6

first disrupted. This can be done by treating the starch with acids, alkalis or enzymes but by far the most common method is by boiling in water (see below).

Another very important chemical related to carbohydrate is ascorbic acid (vitamin C) which will be mentioned later.

Some important reactions of carbohydrates

Fermentation. Originally the term fermentation referred to the foam production during the fermentation of European beer and wine ('fermentare' means 'to cause to rise' in Latin). In the 19th century the great French chemist Pasteur (1822–95) used the term for the production of alcohol and other products by yeast in the absence of air. Today the term is used for enzyme reactions by means of which molecules, usually sugars, are broken down without the aid of oxygen or oxidised inorganic molecules.

In the food industry many types of fermentation are carried out. Examples are the production of alcohol and acids in the manufacture of traditional beers, the fermentation of cocoa, tea, soya, eggs, fish and milk products as well as those processes in which the fermenting microorganism is harvested (see Chapter 14).

The effect of heat. When starch is heated in excess water, the granules hydrate and swell. Their crystalline structure breaks down and under the microscope they appear as bags filled with a colloidal starch solution (Fig. 2.8a, b). The viscosity of the solution increases and this causes the thickening of gravies and sauces when heated. If water is limited as in bread dough, swelling is restricted and the starch granules

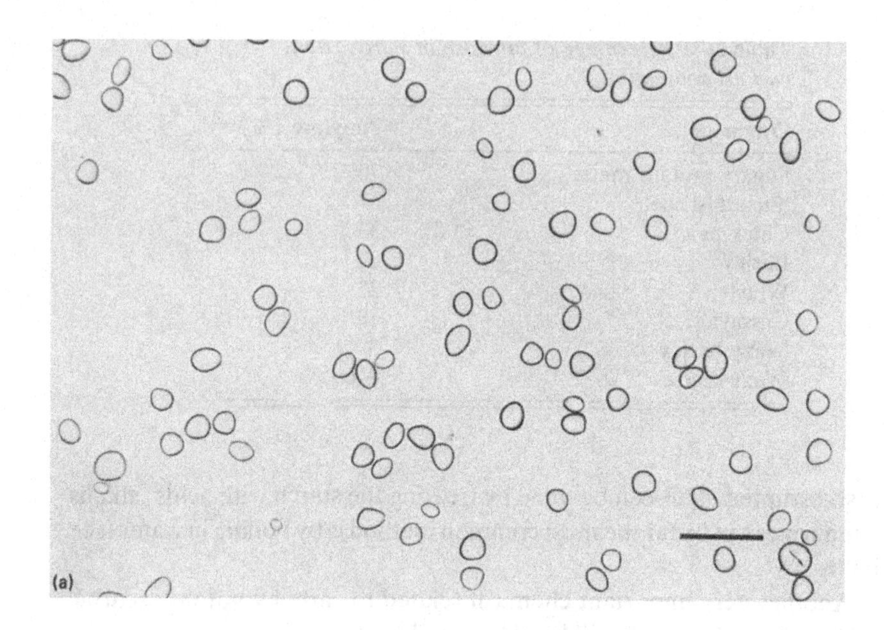
(a)

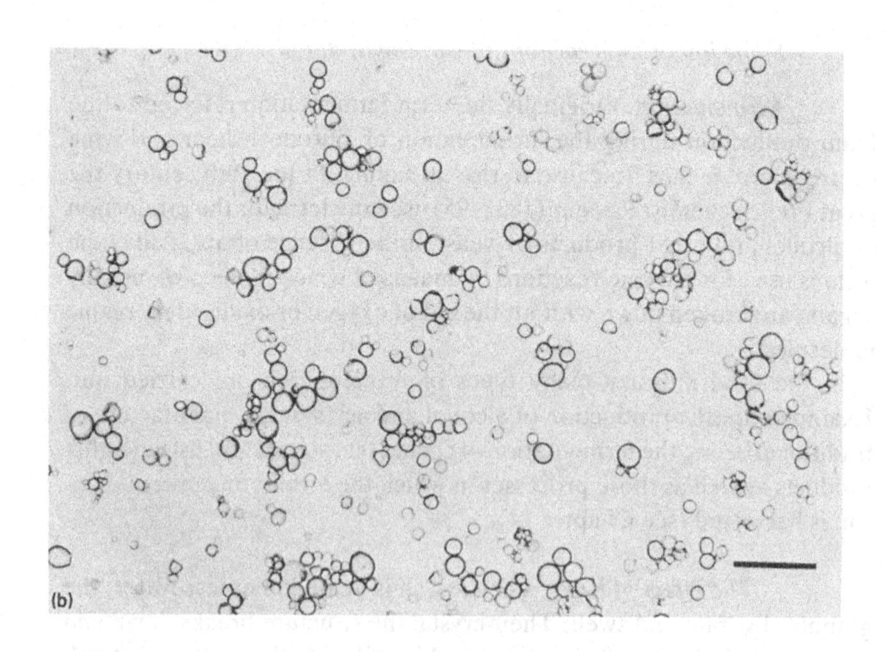
(b)

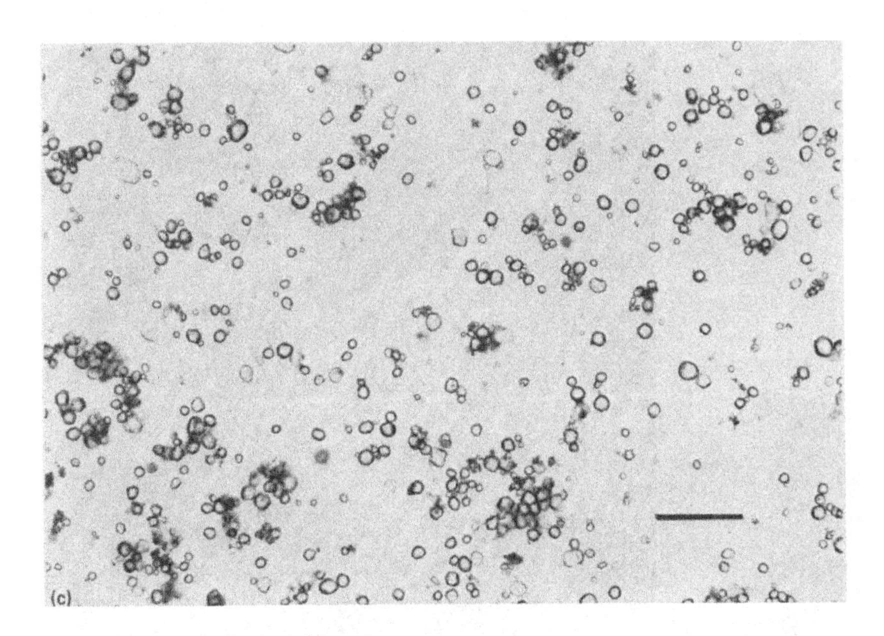

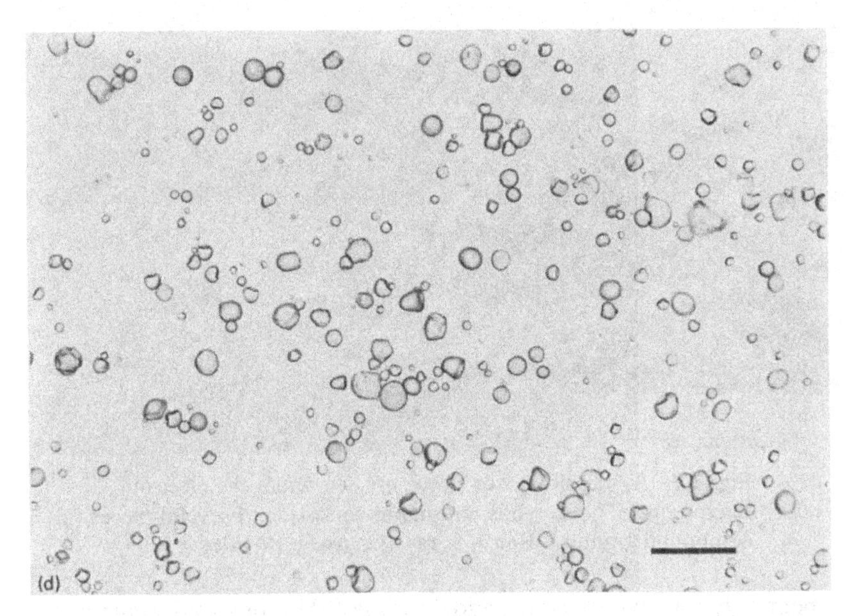

Fig. 2.6a–d. Photomicrographs of starch; bar = 100 μm. (a) yam; (b) cassava; (c) maize; (d) sweet potato.

Fig. 2.7. A granule of rice starch as seen under the electron microscope. The starch is unusual in consisting of several pieces (a compound granule). Bar = 1 μm. (Courtesy R. Reed.)

appear only somewhat swollen and deformed. This process is referred to as gelatinisation. After some time the starch molecules pack together again and recrystallise. This is referred to as retrogradation. It is the cause of bread staling and the leakage of water from starch gels. The latter is called 'bleeding' or 'syneresis'.

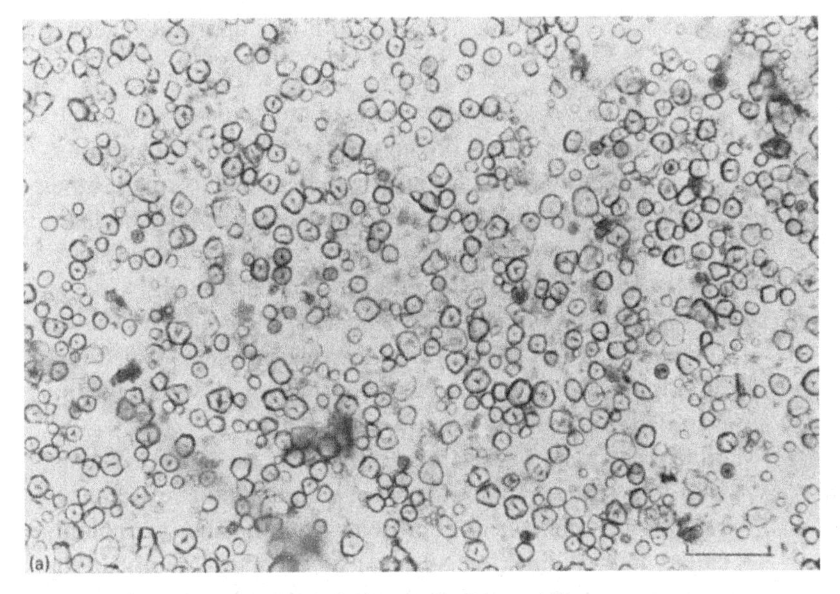

Fig. 2.8a. Raw sorghum starch. Bar = 100 μm.

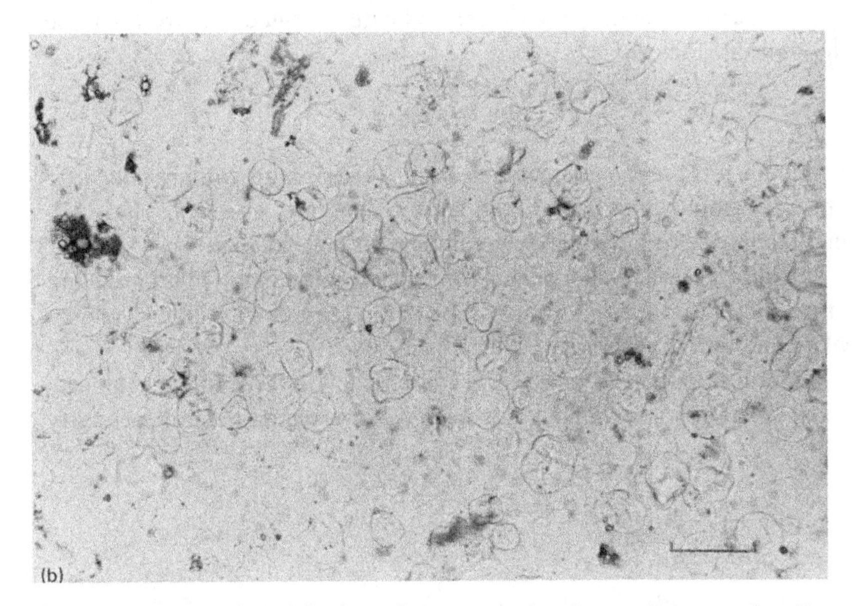

Fig. 2.8b Gelatinised sorghum starch showing swollen granules. Bar = 100 μm.

Non-enzymatic browning. There are three important reactions of this type.

(1) Caramelisation. When sucrose is heated to about 135°C it melts and turns brown. This brown product has a pleasant bitter-sweet taste and is used for making sweets and to colour other food products such as soft drinks, beers and grocery products. If heating is continued, the sugar becomes darker and more bitter and eventually very pure carbon is obtained.

(2) Ascorbic acid browning. When heated to about 40°C in air ascorbic acid (vitamin C) forms brown products which may discolour fruit juices particularly orange juice. Such browning is inhibited by careful exclusion of oxygen.

(3) Maillard reactions. These are so called after the French chemist who first studied them. They are combinations of reducing sugars with amino acids or proteins resulting in brown colours called melanoidins ('melas' means 'brown' in Greek). In the reactions which depend on temperature, pH and concentration some essential amino acids are destroyed. This is particularly serious when intravenous dripfeeds containing sugars and proteins are sterilised in hospitals.

The nutritional value of carbohydrates is not decreased by heat, light, acid or alkali. Indeed, starch must be heated (boiled) before it can be digested satisfactorily.

Fats

In animals fats are the main energy stores. Most plants contain only little fat, apart from some that contain a great deal in their seeds. These oil-seed plants are commercially very important (see Chapter 8).

The difference between oil and fat is not very distinct. Oil is liquid and fat is solid at 'room temperature'. So a fat in temperate regions could be called an oil in the tropics.

Fat molecules consist of glycerol combined with three long-chain fatty acids. On combination one molecule of water is eliminated for each fatty acid.

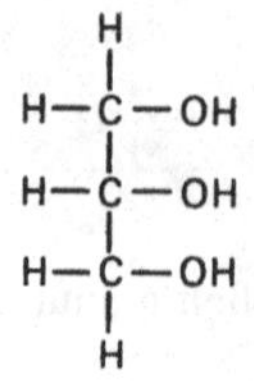

```
     H
     |
 H—C—OH
     |
 H—C—OH
     |
 H—C—OH
     |
     H
```

Glycerol

Each fatty acid consists of a chain of 4–18 carbon atoms. If there are no double bonds the fatty acid is referred to as saturated but if there are double bonds it is an unsaturated fatty acid. Mono-unsaturated fatty acids have one double bond. Poly-unsaturated fatty acids have several double bonds.

Some common fatty acids are shown below.

Palmitic acid (saturated)

```
  H H H H H H H H H H H H H H H   O
  | | | | | | | | | | | | | | |  //
H-C-C-C-C-C-C-C-C-C-C-C-C-C-C-C-C
  | | | | | | | | | | | | | | |  \
  H H H H H H H H H H H H H H H   OH
```

Oleic acid (mono-unsaturated)

```
  H H H H H H H H H H H H H H H H H   O
  | | | | | | | | | | | | | | | | |  //
H-C-C-C-C-C-C-C-C-C=C-C-C-C-C-C-C-C-C
  | | | | | | | |     | | | | | | |  \
  H H H H H H H H     H H H H H H H   OH
```

Linoleic acid (poly-unsaturated)

```
  H H H H H H H H H H H H H H H H H   O
  | | | | | | | | | | | | | | | | |  //
H-C-C-C-C-C-C=C-C-C=C-C-C-C-C-C-C-C-C
  | | | | |     |     | | | | | | |  \
  H H H H H     H     H H H H H H H   OH
```

Linolenic acid (poly-unsaturated)

```
  H H H H H H H H H H H H H H H H H   O
  | | | | | | | | | | | | | | | | |  //
H-C-C-C=C-C-C=C-C-C=C-C-C-C-C-C-C-C-C
  | |     |     |     | | | | | | |  \
  H H     H     H     H H H H H H H   OH
```

Table 2.8 gives the names of some common fatty acids with the numbers of their carbon atoms and double bonds. Linoleic and arachidonic acids are essential to man and may play a role in the symptoms of protein-energy malnutrition (p. 294). Well documented essential fatty acid (EFA) deficiency in man is very rare.

Determination of fats

Qualitatively the fats can be demonstrated by placing a drop of fat stain on the plant or animal tissue and then washing it off. If the tissue contains fat or oil the colour is not easily removed from the fat. Sudan III for instance will stain fatty materials red. Such fatty materials are called lipids

Table 2.8 *Some common fatty acids and the numbers of their carbon atoms and double bonds*

Name	Carbon atoms	Double bonds
Saturated fatty acids		
Butyric	4	0
Palmitic	16	0
Stearic	18	0
Mono-unsaturated fatty acids		
Oleic	18	1
Erucic	22	1
Poly-unsaturated fatty acids		
Linoleic	18	2
Linolenic	18	3
Arachidonic	20	4

and include, apart from fats and oils, also waxes and fatty materials containing phosphorus (phospholipids) or sugars (glycolipids). There are also protein-bound lipids and a chemically complicated material called cholesterol. Cholesterol is found only in animals and not in plants. Egg yolk and dairy products are particularly rich sources. It has been suggested that in some developed countries high cholesterol levels in the blood are related to a high incidence of heart disease.

Quantitatively fats are determined by extracting the material with ether. Some food is weighed and dried. It is placed into a cellulose thimble and continuously extracted with petroleum ether in a Soxhlet apparatus (Fig. 2.9). An ingenious siphoning system empties the barrel at regular intervals and washs the extracted fat into the flask below. After 16 h the flask is disconnected and the ether evaporated. The flask is weighed before and after the test and the difference gives the weight of the fat.

During the past 20 years gas–liquid chromatography (GLC) has been developed to such an extent that it can now be used for the routine analysis of many organic compounds including fatty acids. The methyl esters of the fatty acids are prepared and volatilised. They are then passed at a constant flow rate through a tube filled with a powder (the 'column'). The rate at which the esters travel depends on their affinity with the column. In this way the fatty acid esters are separated and can be detected as they emerge from the column. Figure 2.9a shows the methyl esters of fatty acids (shown as acids) obtained on hydrolysis of margarine.

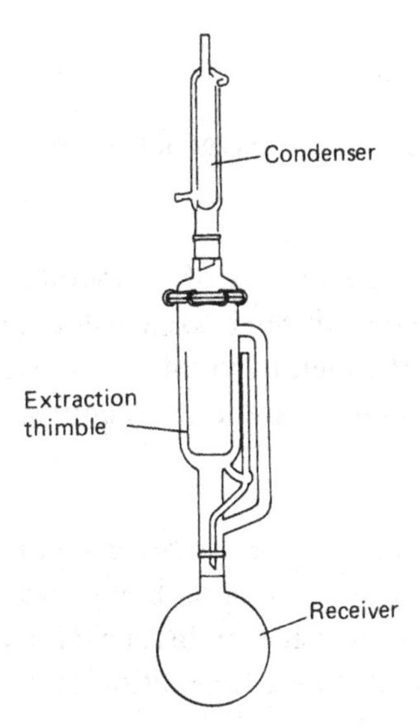

Fig. 2.9. The Soxhlet Extraction apparatus for the determination of ether extract.

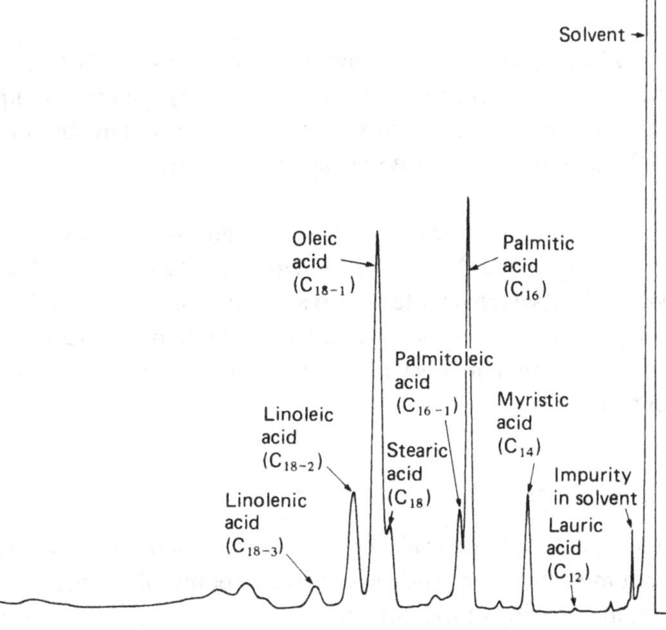

Fig. 2.9a. Methyl esters of fatty acids (shown as acids) obtained upon hydrolysis of margarine.

Some important reactions of fats

Important reactions of fats are hydrogenation, oxygenation, hydrolysis and emulsification.

Hydrogenation. Under proper conditions of temperature and pressure two atoms of hydrogen can be introduced into each double bond of unsaturated fatty acids. In this way the melting point is increased and oil is converted into fat. This process is important in the manufacture of margarine.

Oxygenation. When unsaturated fatty acids are kept in contact with air or oxygen for some time rancidity occurs. Light and warm temperatures accelerate the process. Fats which contain rancid fatty acids are generally unappetising and sometimes toxic. In addition destruction of vitamins A and E may occur (see later).

The very simplified reaction is shown below.

```
 H   H   H   H             H   O        O   H
 |   |   |   |             |   ||       ||  |
-C — C = C — C —   ——→   — C — C   +    C — C —
 |           |             |   |        |   |
 H           H             H   OH       OH  H
```

Hydrolysis. The most important hydrolysis of fats occurs in the digestive system where fats are broken down by the enzyme lipase into glycerol and fatty acids. Hydrolysis can also be brought about by boiling fat in alkali. This takes place in soap manufacture.

Emulsification. Most fats do not mix readily with water. One method of improving this is by dispersing the fat as very small droplets in the water. This system is referred to as an emulsion. A typical example is milk (Fig. 2.10). Emulsions are not very stable and unless emulsifying agents are used two phases will separate. This occurs in the separation of cream from milk.

Vitamins

Like the essential amino acids the vitamins cannot be manufactured in the body but must be eaten ready made. A few points about the vitamins are worth remembering at the outset.

(1) The name 'vitamin' means nothing. Vitamins were first thought

Fig. 2.10. Cow's milk showing fat droplets. Bar = 100 μm.

to be amines, a group of nitrogen containing substances and were called 'vital amines' or 'vitamines'. When it was found that they were not amines at all it was too late to change the name.

(2) There is a family relationship amongst the carbohydrates, amongst the fats and amongst the proteins. Chemical analysis makes it clear to which group these nutrients belong. That is no so with the vitamins: they are all unrelated organic chemicals.

(3) All vitamins are essential to life. If they are not provided in the diet in sufficient amounts, growth is retarded and there is ill-health. If only one vitamin is in extremely short supply, serious illness and finally death are inevitable. Fortunately many vitamins are so widely distributed in the diet, that we would starve to death long before we would be short of that vitamin. Nevertheless some vitamins *are* sometimes deficient and this can lead to very serious consequences. Vitamin deficiency diseases of importance in the tropics are considered in Chapter 17.

(4) Some vitamins are water-soluble and any excess is excreted in the urine. Some are fat-soluble, are not readily excreted and excess may lead to poisoning (hypervitaminosis). There is normally no danger of hypervitaminosis from food intake but it can arise when too many vitamin preparations are eaten.

(5) Many vitamins have three names: (1) a letter, e.g. Vitamin A; (2) its chemical name, e.g. retinol; and (3) a name based on the disease it prevents, e.g. antixerophthalmic vitamin, i.e. it prevents xerophthalmia.

(6) Sometimes the vitamin itself need not be eaten but a closely related substance can be utilised. If we eat that substance the body can transform it into the vitamin with a few simple changes. Such a substance is called a provitamin or a precursor (from pre-currere meaning 'running before' in Latin).

(7) Many vitamins are destroyed during food preparation: by leaching, heating, light, acids or alkalis.

It now remains to list the vitamins very briefly. If they can cause serious disease, they are further considered in Chapter 17.

Vitamin A (retinol)

Fat-soluble, found in animals especially the liver. The precursor called carotene (see below) is found in plants. It is easily destroyed by ultraviolet light, acid, heat or exposure to air (oxygen), but can be protected by antioxidants (i.e. substances that prevent oxidation).

Table 2.9 *Sources of vitamin A and carotene (μg/100g)*

Material	Vitamin A	Carotene (Vitamin A equiv.)
Beef	0	—
Canned sardines	0	—
Fish oil	20–900 × 10^3	—
Butter	1000	—
Margarine	900	—
Eggs	140	—
Cereals (except:)		Trace
Yellow maize		200
Avocado		500
Plaintain		800
Mango		3000
Dark-green leaves (amaranth)		6000
Pale-green leaves (cabbage)		2000
Red palm oil		12000

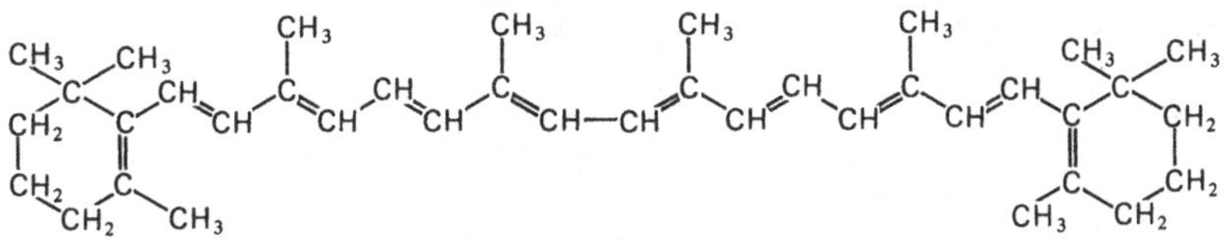

β-carotene

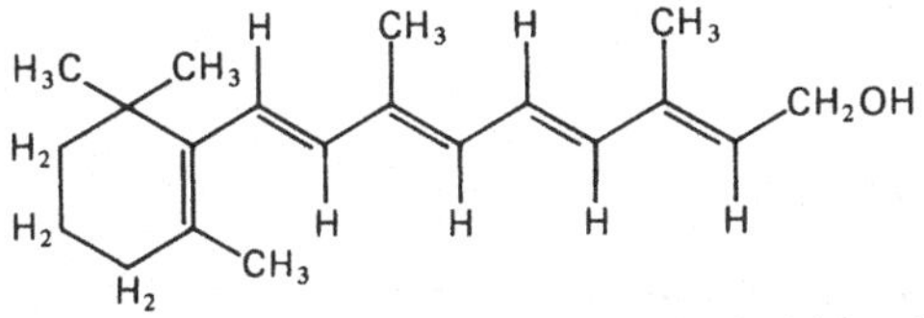

Retinol (vitamin A_1)

Table 2.9 gives some sources of Vitamin A, and the carotene equivalent to vitamin A in μg/100g; 6 μg of β-carotene have the biological activity of 1 μg of retinol.

Vitamin D

Fat-soluble. Only small amounts occur in animals, mainly in fish liver. A precursor called ergosterol is found in plants and is converted to vitamin D by ultraviolet light (sunshine) in the human skin.

Table 2.10 *Sources of vitamin D (μg/100 g)*

Evaporated milk (enriched)	3.0
Cheese	0.3
Eggs	2.0
Liver	1.0
Canned sardines	8.0
Herring	23.0
Butter	0.8
Margarine (enriched)	8.0
Dry Ovaltine	30.0
Fish liver oil	72.0

The structures of vitamin D

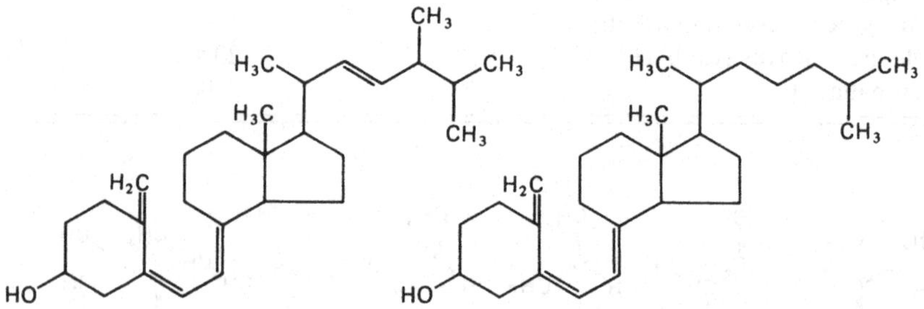

Vitamin D_2 Vitamin D_3

Unstable at alkaline pH and broken down by oxygen, light and heat. Table 2.10 gives some sources of vitamin D.

Vitamin E (tocopherol)

Fat-soluble, most foodstuffs contain small amounts but oil seeds are particularly rich sources. Like vitamin A, tocopherol is easily oxidised and destroyed by ultraviolet light. It is resistant to heat up to 200°C.

The structure of vitamin E (the α-form)

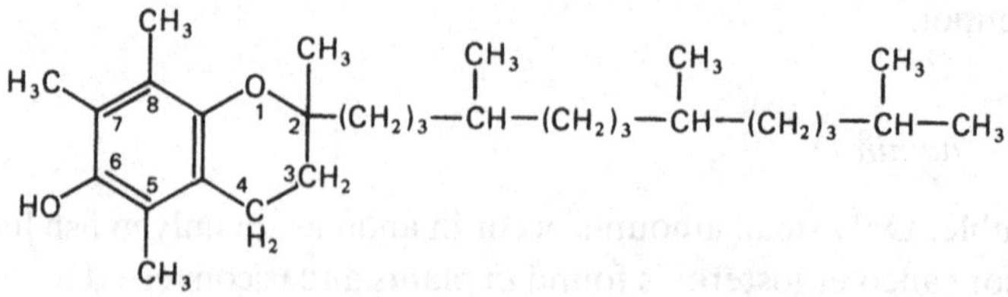

α-Tocopherol

Because it is so readily oxidised it is often used industrially as an antioxidant to mop up oxygen and so protect other materials.

Vitamin K

This consists really of a group of chemicals, all water-insoluble unless specially prepared. The vitamin is found in green vegetables and also is produced by bacteria in the human digestive tract. It is destroyed by oxidation, reduction and ultraviolet light.

The structures of vitamin K

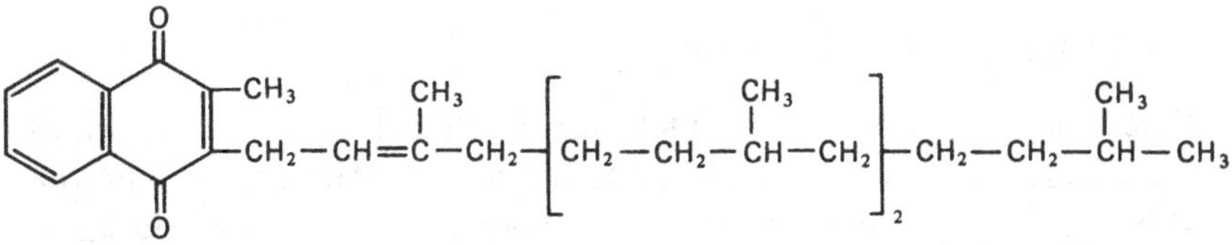

$$-CH_3$$
$$-CH_2-CH{=}\underset{}{\overset{CH_3}{\overset{|}{C}}}-CH_2-\left[CH_2-CH_2-\overset{CH_3}{\overset{|}{C}H}-CH_2\right]_2-CH_2-CH_2-\overset{CH_3}{\overset{|}{C}H}-CH_3$$

Vitamin K_1

$$CH_3$$
$$\left[CH_2-CH{=}\overset{CH_3}{\overset{|}{C}}-CH_2\right]_n-H$$

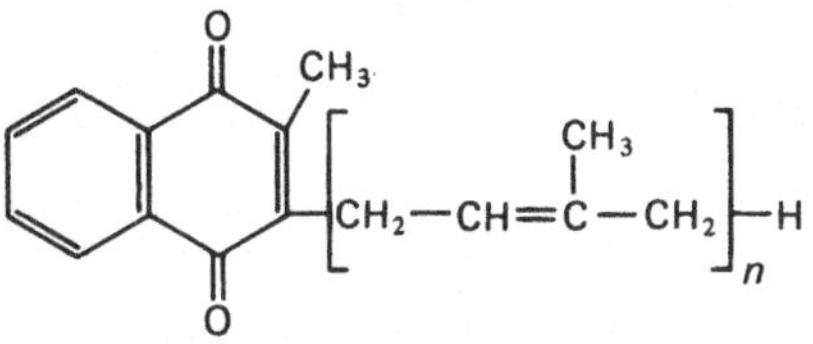

Vitamin K_2

(where n may be 6,7,8,9, or 10)

Vitamin B_1 (thiamin)

This vitamin is water-soluble and present in most foods. Cereals are a particularly good source. Thiamin is quite stable under acid conditions but destroyed when exposed to high temperature for long. In neutral or alkaline solution it is much less stable and destroyed by boiling even for a short time. The vitamin is also destroyed by oxidation, reduction and sulphur dioxide which is sometimes used as a preservative. Table 2.11 gives some sources of thiamin.

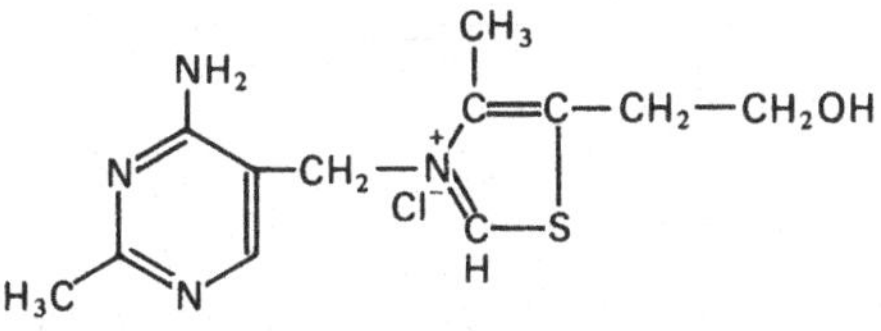

NH_2, CH_3, $C{=}C-CH_2-CH_2OH$, N, $-CH_2-\overset{+}{N}$, Cl^-, C—S, H, H_3C, N

Vitamin B_1

Thiamine

Table 2.11 *Sources of vitamin B_1 (thiamin) (mg/100 g)*

Cereals	0.2–0.3
Roots and tubers	<0.1
Legumes	0.1–0.8
Fats	0
Fish	0.05–0.1
Lake fly	1.0
Meat	0.1–0.3
Liver, kidney	0.2–0.3

Vitamin B_2 (riboflavin)

Water-soluble, the vitamin is widely distributed in leafy vegetables, most meats and fish (Table 2.12). It is stable in acid but not in alkaline solution. Although decomposed by light and ultraviolet light it is relatively stable to heat below 280°C.

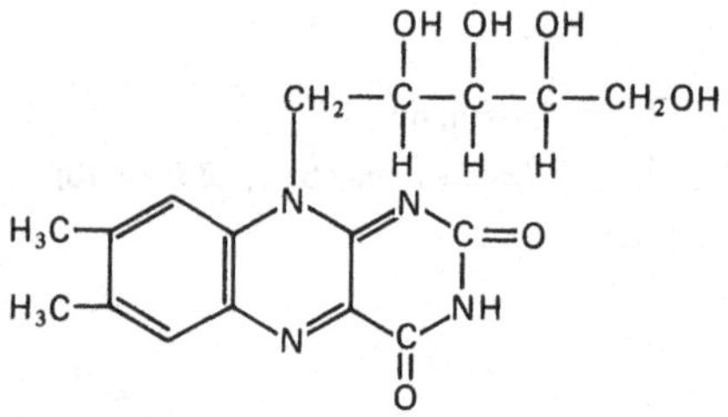

Vitamin B_2
Riboflavin

Table 2.12 *Sources of vitamin B_2 (riboflavin) (mg/100 g)*

Kidney	2.0	Onion, okra	0.1
Liver	2.5	Beans	0.1–0.3
Lake fly	3.5	Roots and Tubers	0.02–0.1
Cheese	0.3–0.5	Cereals	0.05–0.1
Yeast	2.0–3.0		

Vitamin B_5 (pantothenic acid)

This is sometimes called vitamin B_3. It is widely distributed in plants and animals, and liver is a good source. It is quite stable and soluble in water but destroyed by acids, alkalis or heat.

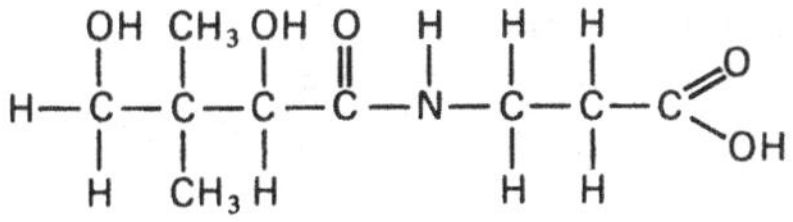

Pantothenic acid

Vitamin B_6 (pyridoxine)

This occurs in most animal and plant cells. Fish, eggs and cereal germ are particularly rich sources. It is very soluble in water and stable to heat even in acid or alkaline solution. It is, however, destroyed by strong alkalis or ultraviolet light.

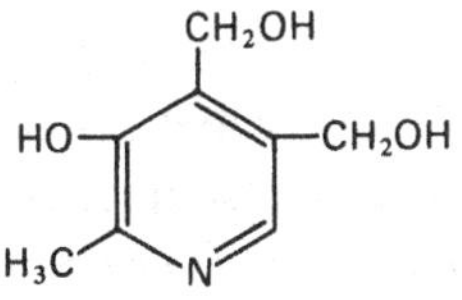

Pyridoxine 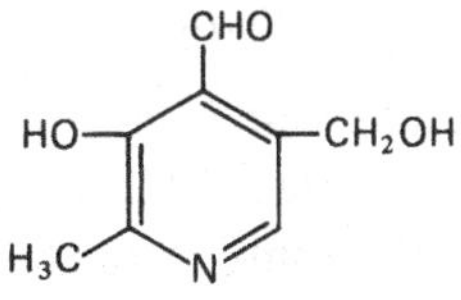 Pyridoxal 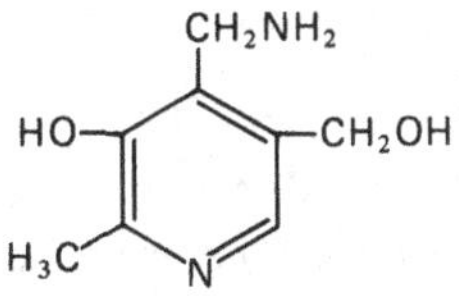 Pyridoxamine

Vitamin B_6

Nicotinic acid (niacin)

This is found in most foodstuffs but in some, e.g. in maize, it occurs in a nutritionally unavailable form. It is made available by boiling the maize in lime water. For this reason the deficiency disease pellagra was common in maize-eating areas of Europe but rare in South America where maize is steeped in lime water before being milled.

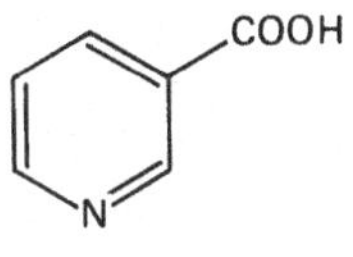

Nicotinic acid 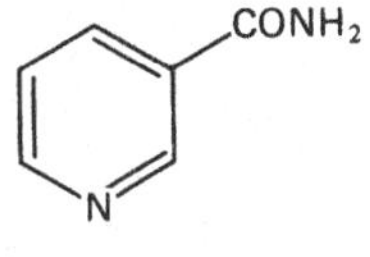 Nicotinamide

Nicotinic acid can also be produced in the liver from the amino acid tryptophan. The vitamin is fairly soluble in water and very stable to heat, oxidation, light or alkali. Table 2.13 gives some sources of niacin.

Table 2.13 *Sources of niacin (mg/100 g)*

Groundnut	10
Baker's yeast	10
Dried yeast	35
Lake fly	18
Liver	13
Mushrooms	6
Most fruit and vegetables	<1
Rabbit	10
Most animal products	0–5
Most roots and tubers	<1
Most cereals	1–5

Vitamin B_{12} (cyanocobalamin)

This is a very large molecule containing one atom of cobalt. It is almost entirely absent from plant material but liver is a rich source. It is also produced by some microorganisms during fermentation. Vegetarians are sometimes at risk from vitamin B_{12} deficiency.

Vitamin B_{12} Cyanocobalamin

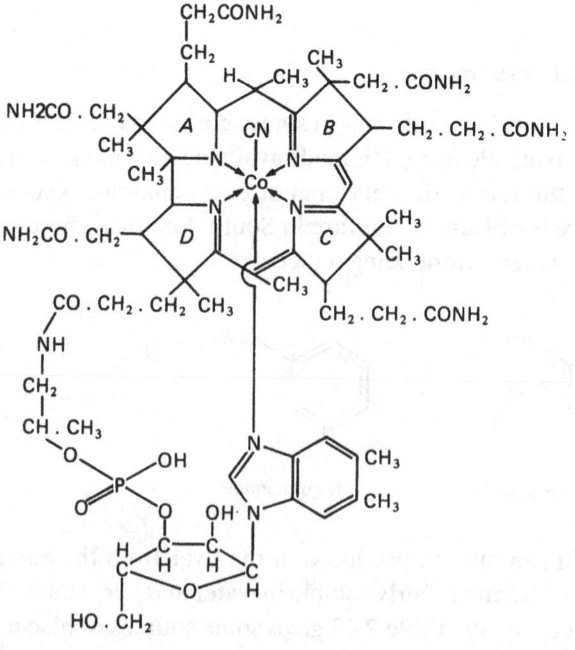

The vitamin is very soluble in water but easily destroyed by natural and ultraviolet light, oxygen and to a lesser extent by some acids and alkalis.

Vitamin M (folacin, folic acid)

This is found in green leafed vegetables and liver. It is only slightly soluble in water and stable to heat in alkaline solution. It is much less stable in neutral or acid solution and is destroyed by light and oxygen.

Vitamin M Folic acid

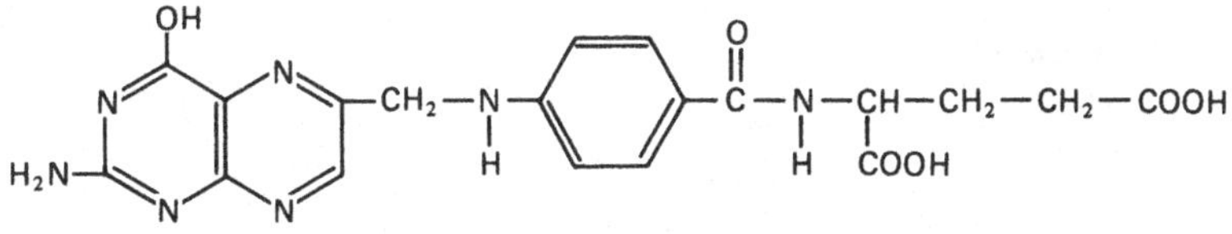

Vitamin C (ascorbic acid)

This vitamin is related to a carbohydrate with the formula shown below.

Vitamin C

Ascorbic acid

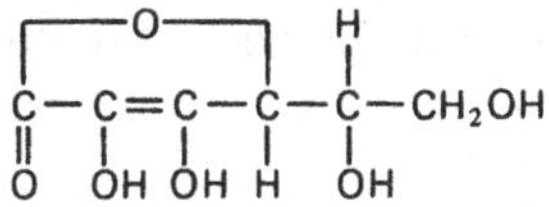

Table 2.14 *Sources of vitamin C (ascorbic acid) (mg/100 g)*

Locust bean (yellow pulp)	200
Baobab juice	75
Baobab dry pulp	350
Soya bean sprouts	10
Guinea sorrel	15
Fresh cassava	30
Plaintain	20
English potato	15
Yam	10
Bean sprouts (*Phaseolus* spp.)	30–100
Sweet peppers	150

It is widely distributed in fruits and green vegetables. Fruits, particularly citrus fruits and cashew apple are good sources. Leaves, particularly dark green leaves (spinach, sweet potato), are a good source. Vitamin C is absent from cereals or legumes unless they are sprouted. Vitamin C content is very variable and may differ from one sample to the next by a factor of 35 due to genetic influence. It is very soluble in water and easily leached or destroyed by heating. Losses on cooking or blanching (p. 249) may be quite high. Table 2.14 gives some sources of vitamin C.

Vitamin H (biotin)

This is obtained mainly from liver and kidney. It is sparingly soluble in water but very stable to dilute acids and alkalis, heat or light.

Vitamin H
Biotin

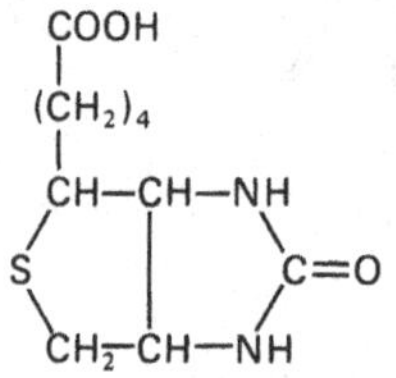

```
        COOH
         |
       (CH2)4
         |
        CH—CH—NH
       /   |    \
      S    |     C=O
       \   |    /
       CH2—CH—NH
```

Minerals

The mineral elements required by the body have three main functions. They form part of the skeleton i.e. bones and teeth, they stabilise body fluids both inside and outside the cells and they are part of enzymes or their cofactors.

Some minerals are found in the body in quite large amounts, i.e. in a matter of grams. Examples are calcium (1200) g), phosphorus (700g), potassium and sulphur (about 200 g), sodium and chlorine (90 g), magnesium (30 g) and iron (2–4 g). These minerals are called major elements, or macro-nutrient minerals.

Other minerals are found only in very small quantities, i.e. in a matter of milligrams. Examples are iodine (20–50 mg), copper (100 mg), manganese (200 mg) and zinc (1–2 mg). Some of these trace elements are never in short supply because the food we eat, whatever the diet, contains ample amounts. It is known that the elements are required only because they form parts of vital enzymes which have been analysed. The body also contains other elements like mercury, gold and lead. No-one knows yet if these are essential. Certainly mercury and lead are very poisonous even in very small amounts.

The mineral elements are all unaffected by light, heat, oxygen tension or pH. They may however be lost on leaching into cooking water, or on peeling vegetables or removal of germ and bran from cereals during the milling process. The ash contents of foods give a good indication of their mineral content. Some of the nutritionally important minerals are discussed below.

Calcium

The skeleton consists largely of calcium phosphate and almost all the body calcium is found in the skeleton.

Meat, root crops and most cereals are poor sources of calcium although some millets contain considerable amounts. In western countries most of the dietary calcium is obtained from milk and cheese. In the tropics other important sources are fish containing edible bones (dried fish, sardines) and in West Africa the giant snail. Caterpillars and some spices are also good sources but they are not eaten that often. In some areas a large proportion of the daily calcium is obtained from the drinking water. Chalk, which is added to white bread in many countries, is a good source.

Only about 20% of all calcium eaten during the day is absorbed and the rest is excreted in the faeces. Some protein in the gut and vitamin D are essential for calcium uptake but phytic acid and oxalate prevent its absorption. These form insoluble salts with calcium as well as with iron and zinc. Phytic acid is common in the bran of cereals and oxalic acid is found in some fruits and vegetables.

The structure of phytic acid is shown below.

Phytic acid

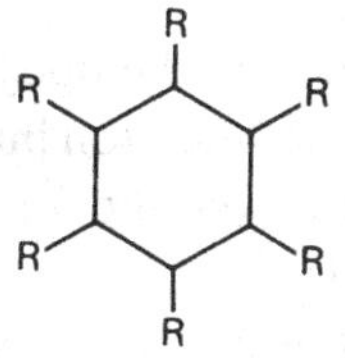

R R R R R R

$R = OPO_3H_2$

Some sources of calcium are given in Table 2.15.

Table 2.15 *Some sources of dietary calcium (mg/100 g)*

Dried fish	3000
Snails	1500
Canned sardines	400
Dried caterpillars	270
Millets	350–400
Cow's, goat's milk	120–130
Skim milk powder	1300
Sesame seeds	1500
Cumin seeds	1000
Meat	5–10

Iron

The body contains about 2–4 g of iron about half of which is found in the red blood pigment haemoglobin. The rest is found in muscle protein, bone marrow and spleen. Some iron is stored in the liver. The daily intake is about 14 mg but only a small portion, about 1 mg, is absorbed.

Various factors play a part in absorption. For instance, iron from animal sources if better absorbed than that from plant sources. If the body is low in iron content the proportion absorbed from the food increases. Ferrous iron is absorbed but not the ferric form and therefore reducing agents such as ascorbic acid and the amino acid cysteine assist absorption. Absorption is inhibited by phytic acid and low iron levels cause iron-deficiency anaemia (see Chapter 17).

Iron is widely distributed in plants and animals but concentration varies so widely that the figures in Table 2.16 are only very rough approximations.

On the whole, cereals and fruit are very poor sources while mushrooms, dried yeast and insects are good ones. Some spices are also useful. Particularly fermented locust bean, sometimes used for cooking, has a high iron content.

Phosphorus

This is the second most abundant element in the body (600–900 g). It is mainly found combined with calcium in the skeleton but is also important in the liberation and utilisation of energy from the food and is contained in some proteins, carbohydrates and fats.

Because phosphorus is essential to all life and contained in practically all foods, deficiency is unknown. The normal intake is about 1.5 g per day.

Magnesium

Magnesium (30 g) is found mainly in the skeleton but also in muscle tissue and various enzymes. Cereals and green vegetables are rich sources: magnesium forms part of the chlorophyll molecule to which the green colour of plants is due. Intake is about 200–300 mg per day and deficiency is rare except when large losses occur during diarrhoea.

Table 2.16 *Some sources of dietary iron (mg/100 g)*

Yam	1	Fermented locust bean	40
Yam flour	10	Cumin, coriander, cinnamon	20–30
Mushrooms	1	Akamu (uncooked)	<1
Dried yeast	20	Liver, kidney	10
Dried lake fly	70	Meat	2–3
Dried caterpillar larva	20		

Sodium and chlorine

Sodium chloride (common salt) is found in all body fluids and serves to maintain the water balance of the body. It is also associated with nerve and muscle activity.

Sodium chloride is low in most foods but is usually added in processing to make the food more palatable. Excess salt is excreted in the urine and in sweat. Intake varies a great deal from 5 to 20 g per day. If the intake is too low, or excretion too high, i.e. due to sweating through working hard in a hot climate, muscular cramp may result.

Salt is easily detected by taste and good sources are salted fish or meat, processed vegetables and margarine. Table salt is virtually pure sodium chloride.

Potassium

Two hundred grams of this element are found mainly within the body cells.

Excess potassium is excreted through the urine but dangerously high losses may result through diarrhoea or in protein-energy malnutrition where tissue breakdown is taking place. If potassium loss is very severe the element must be provided or heart failure may result. The main sources of potassium are vegetables and meat. Fruit and fruit juices may also contain significant amounts of it.

Iodine

The body contains between 20 and 50 mg of iodine, the highest concentration (approximately 8 mg) being found in the thyroid gland. Iodine is particularly important in the production of thyroid hormones which control the metabolic rate of the body. The iodine content of the food depends largely on locality and because the soil is often low in iodine, foodstuffs also often contain little. An exception are sea foods which contain high levels. Iodine may also be made unavailable by cyanide derivatives. These are important in cassava (see Chapter 8). If iodine absorption is too low, goitre may result (see Chapter 17). In many countries of the world goitre has been eradicated by the use of iodised table salt.

Fluorine

Water, with the exception of soft water, is the most important source of fluorine. The element occurs in bones and teeth. It seems to strengthen the teeth and so reduces tooth decay. In many countries fluorine is regularly added to the drinking water providing an intake of 1–2 mg a day. Higher intakes may be dangerous. Apart from the water, important sources of fluorine are tea and fish the bones of which are eaten (fish stew, canned sardines).

Copper

The 100–150 mg of copper contained in the body are associated with several enzymes.

Excessive intake of copper is poisonous and even quite low doses can cause vomiting. Deficiency is rare but has been observed in small children who were fed on little else but cows' milk. This contains less copper than most foods.

Cobalt

The body contains about 2 mg of cobalt. As far as is known its only role is as part of the vitamin B_{12} molecule.

Zinc

One to two milligrams form part of several important enzyme systems and zinc is also found in bone. Most of the zinc ingested is excreted. It is found mainly in high protein foods (meat, eggs, fish) while leafy vegetables and cereals are poor sources. Phytic acid interferes with zinc absorption as it does with that of calcium and iron. Zinc deficiency causes stunted growth.

Manganese

Widely distributed in the body, the element is found mainly in plant products, particularly in cereals, legumes, coffee and above all, tea. Animal sources tend to be low in manganese.

3

Digestion and absorption of food

The alimentary canal

The alimentary canal is a tube running from the mouth to the anus. It is about 9 m long although in a living person it is much shorter (3–5 m) due to muscle contraction. Its purpose is to break down the food, withdraw nutrients from it and expel the residue. Several other organs are associated with the alimentary canal and it is therefore convenient to divide it into six sections as follows.

(1) Mouth, pharynx and oesophagus
(2) Stomach
(3) Duodenum and pancreas
(4) Liver and gall bladder
(5) Small intestine (jejunum and ileum)
(6) Large intestine (colon and rectum)

This is shown in Fig. 3.1.

Except for the very front end (mouth, pharynx, upper part of oesophagus) and the very rear end (external anus) which consist of striated (skeletal) muscle, the alimentary canal consists of smooth muscle. Figure 3.2 shows a transverse section of the small intestine. There is an outer layer of longitudinal muscle and an inner layer of circular muscle. Both the inside and outside of the tube are lined with *epithelium*, the *peritoneum* on the outside and a layer of cells lining the *villi* on the inside. The villi are fingerlike processes typical of the small intestine (Fig. 3.3a, b). They greatly increase the surface of that part of the gut to make absorption of nutrients easier. The villi are well furnished with blood capillaries and lacteals. These are lymph vessels which serve the absorption of fatty materials. The blood vessels absorb monosaccharides and amino acids.

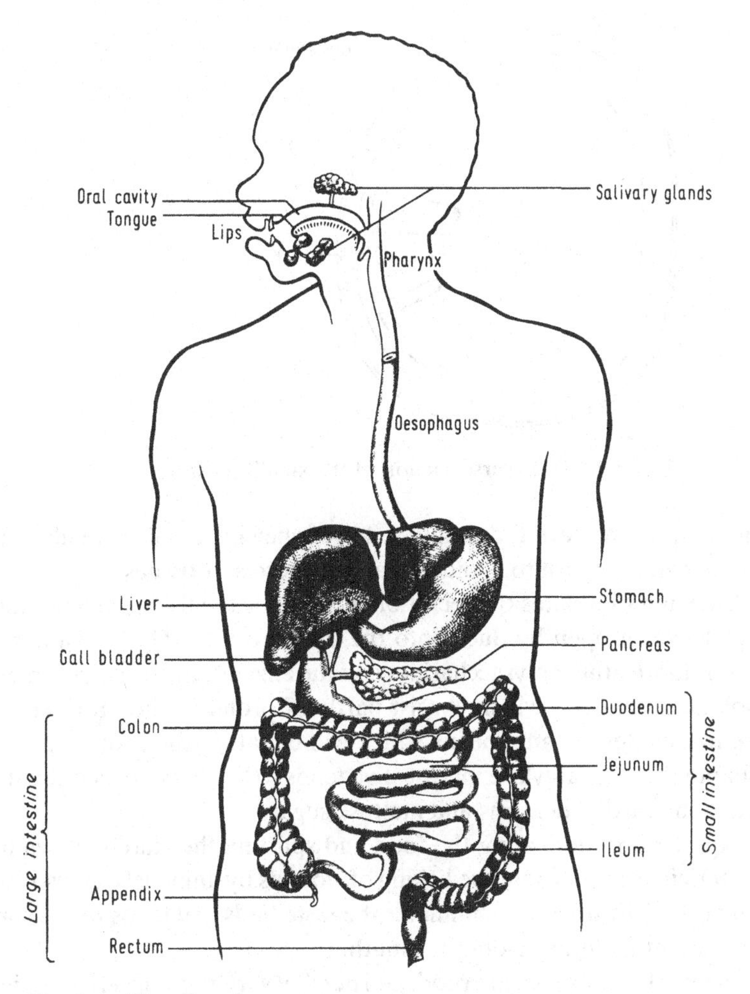

Fig. 3.1. The digestive system.

Apart from blood and lymph vessels there is also a nerve supply to the gut. This affects both movement and secretion of glands and sensory nerve fibres also relay information from the gut.

(1) Mouth, pharynx and oesophagus

The mouth is chiefly concerned with chewing which is either a conscious voluntary act or a reflex action. The front teeth, the incisors, are used to cut the food into suitable pieces while the back teeth or molars grind the food. Chewing forces are very considerable, in the region of 300–1000 Newtons. The tongue and cheeks are used to position the food between

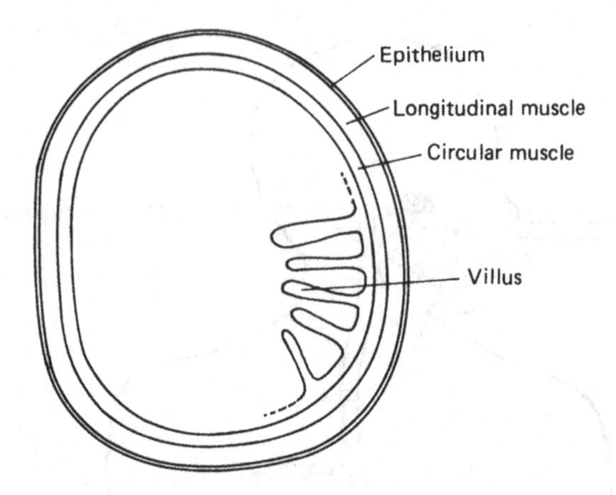

Fig. 3.2. Transverse section of the small intestine.

the teeth. Therefore the simple reflex of chewing involves really a quite complicated control to avoid damage to these soft tissues.

Chewing stimulates the secretion of saliva by the three pairs of salivary glands which open by ducts into the buccal cavity. The saliva contains water, lubricating polysaccharides and the enzyme amylase. Enzymes are biological catalysts which consist partly of protein. Therefore enzymes are inactivated by any means which will denature the protein part (see Chapter 2). The amylase in the saliva tends to break down cooked starch into a mixture of dextrins and maltose sugar.

Apart from lubricating the food and splitting the starch, saliva has a bactericidal effect (note the licking of wounds by animals!). It also allows flavours to dissolve and stimulates the taste buds and helps to wash away unpleasant materials from the mouth.

In man 1.5 l of saliva are produced per day varying from a few millilitres per hour to 250 ml per hour during a meal.

Swallowing is initiated when the tongue pushes a lump (bolus) of food against the first part of the pharynx. This causes a reflex which closes the respiratory tube and pushes the bolus down the oesophagus by peristalsis. When the food reaches the lower part of the oesophagus another reflex allows the food to enter the stomach.

(2) The stomach

The stomach is a muscular bag closed at each end with a sphincter muscle and is covered on the inside with a lining called the gastric mucosa. This is thrown into folds called rugae which tend to disappear when the stomach is full. There are also pits in the mucosa.

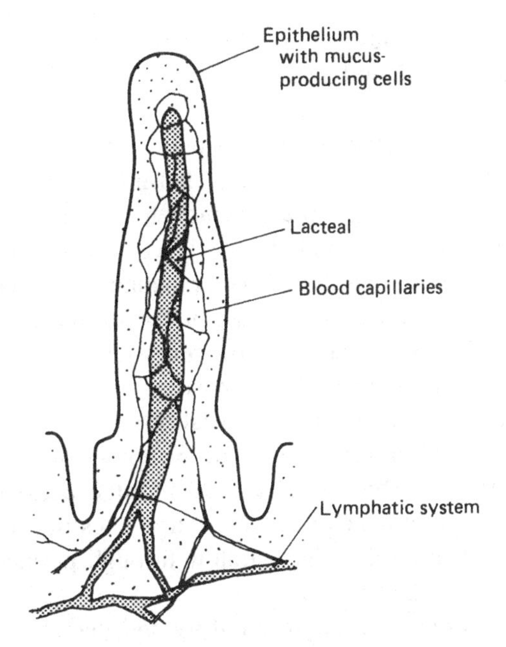

Fig. 3.3a. Structure of a villus.

Fig. 3.b. Scanning electron micrograph of the small intestine of a mouse. At bottom gut wall in section. Above, fairly thick villi in side view. At top, villi in surface view. Bar = 200 μm. (Courtesy R.L. Holmes.)

There are several types of glands in the lining of the stomach which secrete mucus, hydrochloric acid, and inactive enzymes (enzyme precursors) called pepsinogens.

The presence of acid in the stomach tends to inhibit the salivary amylase but itself hydrolyses sucrose. It also kills harmful microorganisms eaten with the food and changes the pepsinogens into the active pepsins. The pepsins split peptide bonds in the ingested protein and convert the large protein molecules into smaller amino acid chains. Because the gastric mucosa secretes the enzyme precursor and not the enzyme itself the stomach is protected from its own enzyme and does not digest itself. The mucus produced also acts as a barrier.

The acid-producing cells also secrete a protein called instrinsic factor. Very little is known about this but it is essential for the uptake of vitamin B_{12}. If this vitamin is not absorbed death results through pernicious anaemia. The removal of the stomach in an operation does not affect digestion very much but death will quickly follow through pernicious anaemia unless intrinsic factor is supplied.

The peristalsis of the stomach serves to emulsify and macerate the contents and eventually lots of about 5 ml of the liquified food called chyme are ejected into the duodenum. In addition to normal peristalsis much more severe contractions of the stomach occur 12–24 h after a meal. These give rise to hunger pains. Alcohol, glucose and water-soluble vitamins may pass through the walls of the stomach, but otherwise there is very little absorption.

(3) Duodenum and pancreas

The duodenum is only about 25 cm long but nevertheless is a major site of digestion. Both the bile duct from the gall bladder and the pancreatic duct run into it. The duodenum itself contains glands which secrete alkaline mucus to protect it from the gastric acid and pepsin.

The pancreas consists of two major types of cell groups. One type discharges digestive enzymes into the gut and the other, the Islets of Langerhans, secrete two hormones (glucogen and insulin) into the bloodstream.

The pancreatic juice flowing into the duodenum consists of an alkaline fluid to neutralise the acid chyme and to provide an alkaline medium in which the pancreatic enzymes work best. These enzymes consist of four which attack proteins, an amylase similar to that in saliva which attacks starch and a lipase which separates the fatty acids from glycerol. So important are these pancreatic enzymes that if the pancreas is surgically removed 15–80% of the protein and 50–60% of the dietary fat are not utilised and are lost in the faeces.

Although a great deal of enzymatic digestion occurs in the duodenum there is very little absorption.

(4) Liver and gall bladder

The liver is the largest gland in the human body. Removed it weighs about 1500 g but in the living body it is considerably heavier because one of its functions is the storage of blood. Another is the breakdown of red blood corpuscles and the removal of iron from the haemoglobin molecule and the storage of iron as ferritin. The liver is the big chemical factory of the body. The blood from the intestine carrying nutrients (see Chapter 4) does not pass directly into the general circulation but is brought first to the liver via the hepatic portal vein. Thus all digested carbohydrate and protein must pass through the liver first. The carbohydrate is released into the blood as sugar in the required amount and any excess is stored as glycogen. The amino acids are also released in suitable quantities but excess is deaminated and excreted by the kidneys as urea.

The gall bladder is a pear-shaped sac 6–10 cm long and lies in a hollow on the surface of the liver. Various ducts bring the bile from the liver to the gall bladder which itself empties into the duodenum. The main secretion consists of bile acids but there are also small quantities of cholesterol and bilirubin which is a breakdown product of the red blood corpuscles. Bilirubin gives bile its golden brown colour, but it turns green in an acid medium. The bile acids which help to emulsify the fats in the gut are later reabsorbed, pass into the liver via the portal vein and are circulated in this way. About 5–10% are lost in the gut due to bacterial action and the loss is made good by the liver. Bile acids have an important function in absorption.

(5) The small intestine (jejunum and ileum)

The internal surface of the small intestine is thrown into circular folds and the whole covered with small fingerlike projections, the villi (Figs. 3.2, 3.3). Between the villi lie intestinal glands which renew the epithelial cells, which are regularly shed. It is in this part of the gut that digestion is completed and most of the absorption of nutrients takes place. Both protein- and carbohydrate-degrading enzymes are secreted from the lining of the small intestine. While a series of contractions push the chyme both backwards and forwards mixing the chyme with the digestive enzymes, the overall movement is towards the large intestine. Monosaccharides produced by digestion are absorbed by an active process rather than simple diffusion within the first 100 cm of the jejunum. Amino acids also pass through the gut wall by various active mechanisms. In the newborn it seems that whole proteins are absorbed and in this way immunity to some diseases can pass from the mother to the suckling infant (see Chapter 4). Absorption of whole proteins probably decreases after infancy but even in adult life the process may not ceas altogether, being responsible for some food allergies.

Fats, glycerol and cholesterol pass across the inner layer of the gut probably by simple diffusion taking the fat-soluble vitamins with them. The water-soluble vitamins apparently also diffuse passively although it is thought that folic acid is actively absorbed.

(6) The large intestine (colon and rectum)

The large intestine is divided into the caecum, ascending, transverse and descending colon, the rectum and the anal canal. The structure of the large intestine is similar to that of the small intestine but there are no circular folds and no villi. In the large intestine the chyme is converted into faeces through absorption of water. Some mucus is secreted to lubricate the gut content and bicarbonate ions to neutralise acids produced by bacteria. Sometimes in ill health irritation of the inner layer of the large intestine stops it absorbing water and salt and causes it instead to secrete them. This results in watery faeces and is called diarrhoea. When material from the colon enters the normally empty rectum the stretching of its wall causes a desire to defecate. Voluntary contraction of the anal sphincter muscle causes the contents of the rectum to be returned to the colon for storage. Eventually the arrival of further faeces produces a much stronger urge to defecate.

The presence of some undigested material in the colon provides ideal conditions for a gut microflora. This is usually beneficial providing a good source of B vitamins and vitamin K. Excessive amounts of carbohydrate may be fermented producing considerable amounts of gas such as methane, hydrogen and carbon dioxide.

The faeces consist normally of 75% water and 25% solid matter. Of the latter, 30% is undigested fibre, 30% bacterial cells from the colon, 10–20% undigested fat and the remainder inorganic material and small amounts of mucus and protein.

4

Nutritional needs through life

In order to live healthily all people require water, protein, energy, essential fatty acids, vitamins and minerals. So the question arises how much of each is necessary. Obviously there is a large variation depending on age, sex, weight, pregnancy and lactation. These factors can be taken into account but a major difficulty is the variation between individuals. Nutritional requirements have been determined on a relatively small number of people in the laboratory but the results must apply to whole populations. Therefore if one would take the average nutritional requirement, half of the population would be overfed while the other half would be in danger of malnutrition. To avoid this, nutritionists do not use 'absolute requirements' but 'recommended intakes'. This will satisfy the need for any particular nutrient of 87–98% of the population. The daily intake of nutrients recommended by WHO is given in Table 4.1.

Other countries or organisations may have slightly different values sometimes affected by political considerations. For instance free school meals may be based on recommended intakes and their assessment may affect public expenditure.

Recommended energy requirements

The energy used for everyday living can be considered under two headings: the Basal Metabolic Rate (BMR) and physical exertion.

The BMR refers to the energy expended when a person is at complete rest i.e. relaxing in an armchair or asleep at a temperature where heat is neither gained nor lost. Energy expenditure measurement in kJ/min is related to lean body weight. Fat is relatively inert and therefore the BMR refers to the metabolic activity in the lean body. Women usually contain somewhat more fat than men and therefore their BMR for the same

Table 4.1 *World Health Organization Recommended Intakes*

Age	Body weight (kg)	Energy (kcal)	Energy (MJ)	Protein[a] (g)	Vitamin A[b] (μg)	Vitamin D[c] (μg)	Thiamin (mg)	Riboflavin (mg)	Niacin (mg)	Folic acid (μg)	Vitamin B_{12} (μg)	Ascorbic acid (mg)	Calcium (g)	Iron (mg)
Children														
<1	7.3	820	3.4	14	300	10.0	0.3	0.5	5.4	60	0.3	20	0.5–0.6	5–10
1–3	13.4	1360	5.7	16	250	10.0	0.5	0.8	9.0	100	0.9	20	0.4–0.5	5–10
4–6	20.2	1830	7.6	20	300	10.0	0.7	1.1	12.1	100	1.5	20	0.4–0.5	5–10
7–9	28.1	2190	9.2	25	400	2.5	0.9	1.3	14.5	100	1.5	20	0.4–0.5	5–10
Male adolescents														
10–12	36.9	2600	10.9	30	575	2.5	1.0	1.6	17.2	100	2.0	20	0.6–0.7	5–10
13–15	51.3	2900	12.1	37	725	2.5	1.2	1.7	19.1	200	2.0	30	0.6–0.7	9–18
16–19	62.9	3070	12.8	38	750	2.5	1.2	1.8	20.3	300	2.0	30	0.5–0.6	5–9
Female adolescents														
10–12	38.0	2350	9.8	29	575	2.5	0.9	1.4	15.5	100	2.0	20	0.6–0.7	5–10
13–15	49.9	2490	10.4	31	725	2.5	1.0	1.5	16.4	200	2.0	30	0.6–0.7	12–24
16–19	54.4	2310	9.7	30	750	2.5	0.9	1.4	15.2	200	2.0	30	0.5–0.6	14–28
Adult man (moderately active)	65.0	3000	12.6	37	750	2.5	1.2	1.8	19.8	200	2.0	30	0.4–0.5	5–9
Adult woman (moderately active)	55.0	2200	9.2	29	750	2.5	0.9	1.3	14.5	200	2.0	30	0.4–0.5	14–28
Pregnancy (later half)		+350	+1.5	38	750	10.0	+0.1	+0.2	+2.3	400	3.0	50	1.0–1.2	(d)
Lactation (first 6 months)		+550	+2.3	46	1200	10.0	+0.2	+0.4	+3.7	300	2.5	50	1.0–1.2	(d)

(a) As egg or milk protein. (b) As retinol. (c) As cholecalciferol. (d) Dependent on previous iron status.
Source: Davidson *et al.*, 1985.

Table 4.2 *Values for the basic metabolic rate of adults (watts)*

Men	Women	Fat	Weight (kg)							
		(%)	45	50	55	60	65	70	75	80
Thin		5–	—	68	73	78	83	88	92	97
Average		10–	—	65	70	75	80	85	88	93
Plump	Thin	15–	57	62	67	72	77	82	85	90
Fat	Average	20–	55	58	63	68	73	78	82	85
	Plump	25–	—	55	60	65	70	75	78	83
	Fat	30–	—	—	57	62	67	72	75	80

Source: Davidson *et al.*, 1986.

Table 4.3 *Relation between work load and energy expenditure (kJ/min)*

Work load	Energy expenditure
Very light	10
Light	10–20
Moderate	20–30
Heavy	30–40
Very heavy	40–50
Extremely heavy	50

weight is a little lower. Values for the BMR of adults are given in Table 4.2. How climate or race acts on BMR is at present not quite clear.

The energy expenditure due to physical work is added to the energy required for BMR. Table 4.3 shows that it depends on the work load.

People cannot carry on extremely heavy work for very long periods. They require frequent rests. It is therefore not surprising that much of the daily energy expended is due to the BMR and less-energetic activities such as getting dressed in the morning, going to work, having meals and at leisure. Therefore the difference in energy expended in a day between one person doing very heavy work and another doing very light work is not as great as one might expect. This is shown in Table 4.4.

It has also been shown that energy expenditure in the tropics in the wet season was higher than in the dry season and in cooler uplands higher than in low-lying coastal regions. This would suggest that the effect of climate is due mainly to changes in physical activity. It seems reasonable to reduce recommended energy intake by 5–10% where the mean annual temperature exceeds 25°C.

Table 4.4 *Average daily expenditure of energy by 10 clerks (average age 28.3 years weight 64.6 kg) and by 19 miners (average age 33.6 years weight 65.7 kg) as measured over a whole week*

	Clerks		Miners	
	MJ	kcal	MJ	kcal
Asleep and day-time dozing	2.1	500	2.1	490
Activities at work	3.7	890	7.3	1750
Non-occupational activities and recreations	5.9	1410	5.9	1420
Total	**11.7**	**2800**	**15.3**	**3660**

Source: Davidson *et al.*, 1986

Recommended protein requirements

Protein has to be supplied to the body for two reasons. Firstly to build new protein-containing tissues and secondly to replace protein lost in normal metabolism. At an early age the proportion of protein required for body growth is very high and that required to replace protein lost in urine, faeces and skin cells is low. With increasing age the proportion due to protein replacement becomes more important as growth slows down and at about age 19 years new tissues are no longer laid down as growth ceases altogether. The recommended protein intakes are given in Table 4.1. Because the utilisation of protein by the body depends also on protein quality (see Chapter 2), a correction is made to allow for that.

Recommended vitamin requirements

Our knowledge concerning vitamin requirements is very limited and estimates vary from country to country. Vitamins can be determined in the blood and one regards a level as satisfactory which maintains the person concerned in good health. (Note: after a meal levels may be unduly high.)

Where one can predict high metabolic activity in which a vitamin is involved it seems reasonable to suggest a higher intake of that vitamin. For example thiamin is involved in carbohydrate metabolism and if people live on a high carbohydrate diet they require more thiamin. If there are large losses of water through sweating there may be a greater requirement for the water-soluble vitamins which may be excreted in the sweat.

Some vitamins such as vitamin K and the B-group vitamins are also synthesised in the gut (see Chapter 3) by the resident microorganisms. Therefore if medical treatment (e.g. with antibiotics) destroys these, additional vitamins may be required. The recommended intake of some vitamins is shown in Table 4.1.

Recommended mineral requirements

As for vitamins, minerals can be determined in the blood or by carrying out balance studies. With the latter, intake and output of the element are compared and the intake established which prevents a fall in the blood level of that mineral. There are often physiological mechanisms which change the rate of absorption and excretion of the mineral and so concentration is often kept within quite a narrow range. In hot climates the temperature of the body can only be maintained by sweating and a man engaged in heavy work in the tropics can easily lose 4 l of sweat per day and with it about 14 g of salt. This would lead to serious illness and therefore ample cool water should be supplied and the food should be well salted. Wherever possible shade should be provided and hard physical work should be avoided in the midday heat.

The recommended intakes of calcium and iron are given in Table 4.1.

Having considered briefly the nutritional needs of adults those of special groups must now be dealt with.

Feeding the infant

If breast fed, African children grow as well or better than children in highly developed countries. A mother's milk provides almost all the nutrients required. It is low in iron but the newborn has a supply which should last for 3–6 months; only if it is one of twins or premature may its iron store be low. So the normal infant requires no other food during the first 4–5 months. At the end of that period the child is weaned to provide extra protein, iron and vitamins C and D. Table 4.5 gives the relevant analyses of some cooked plant foods. It is apparent that the child will require extra sources of iron and vitamins C and D.

Extra protein can be obtained from dried milk powder, egg, finely minced meat or fish and legumes. Groundnut soup is very nutritious for the toddler. Moderate quantities of beans are useful but the skin should first be removed or there will be excessive flatulence.

Finely chopped green leaves (amaranth, cassava) can provide additional vitamin C, carotene and iron and the first two can also be provided from fruit. From 6 to 12 months the normal child should gain

Table 4.5 *Composition of various cooked plant foods (per 100 g)*

	Water (g)	Energy (kJ)	Protein (g)	Fat (g)	Iron (mg)	Vit. C (mg)	Vit. D (μg)
Beans, boiled	70	400	8	0.5	2	0	0
Bread, white	35–40	950	8	1	1	0	0
Cassava, boiled	70	520	1	0.1	0.5	0	0
Maize, porridge	50–80	NA	2–5	1–4	0.5	0	0
Potato, boiled	80	450	2	0.1	0.2	10	Tr
Rice, boiled	70	500	3	0.1	0.1	0	0
Pasta, boiled	70	550	4	0.7	0.5	0	0
Tortilla, baked	50	900	5	2	1	0	0

Note: NA = not available, Tr = trace.

between 2 and 3 kg in weight. If at all possible the child should be taken regularly to a clinic where growth charts are prepared to make certain that the infant develops well.

The 'full belly' concept

If one is not trained in nutrition it is very difficult to understand how a child can suffer from protein-deficiency disease when it has always plenty of food. Latham (see Further reading) refers to the 'full belly' concept. He reports that during his work in Tanzania he had to explain to the parents of children suffering from kwashiorkor (see Chapter 17) that the disease was due to a poor diet. The parents often explained that the child was never hungry, that whenever it required food it was given porridge. It always had a 'full belly'. It was hard for them to realise that a porridge made from plantain, yam, cocoyam, cassava or sweet potato does not contain sufficient protein for the child's needs. The protein is simply too dilute. The child can eat the porridge to excess and still not have enough protein.

Often people also do not realise that while adults can get all their nourishment in two meals a day, small children do not have such a large stomach. Infants under one year should be fed six times a day, children from one to three years old, four to five times a day and older children three times a day.

Child nutrition

The young child, like the weaned infant, is in danger of protein, iron and vitamin deficiency even if consuming sufficient energy (see Chapter 17).

When it is 2½ years old the child can eat most of the food of the rest of the family but it tends to dislike peppery or spicy foods. Therefore it tends to pick out low protein, high carbohydrate parts of the meal. Since children must be fed more often than adults these additional meals should have a high protein content. At table young children should have the highest priority, before the men. They should have the most nutritious portion of the family meal because their protein requirements are relatively the highest.

When the child is older the midday school meal is ideal for the introduction of new foods. The meals should be designed by someone with a training in dietetics.

Nutrition of the adolescent

As can be seen from Table 4.1 there is a slowly decreasing need for calcium as the skeleton is laid down. However the most important factor is the burst of growth occurring in males at puberty resulting in additional requirements of nutrients. In females this is less pronounced but here the onset of menstruation makes additional intake of iron necessary. Iron-deficiency anaemia is perhaps the most common disorder among women the world over (see Chapter 17). The increased requirement for energy is often satisfied by eating between meals. In highly advanced countries this often leads to high intakes of sweets resulting in bad dietary habits (obesity) and an increase in tooth decay (caries). The sucrose remaining on the teeth is utilised by bacteria and converted into acid which attacks the enamel. In the tropics useful high-energy snacks are roasted maize, coconuts and roasted groundnuts. Fruit is useful for its vitamin content and cold cooked meat or fish or hard-boiled eggs provide important sources of protein.

Pregnancy and lactation

An increased amount of energy is required during pregnancy for the increase in tissues of both foetus and placenta. Fat reserves are laid down by the woman for later milk production and more energy is required to carry the increased body weight. The energy cost of a pregnancy is about 335 MJ of which 150 MJ are associated with fat deposition.

In many rural areas of the tropics women remain active throughout their pregnancy and the whole of their energy needs must be met through the diet. The developing foetus is truly a little parasite and will take calcium, iron, protein and vitamins, particularly vitamin A, from the mother. These nutrients must be supplied in greater amounts than at any time during her life. The health of the mother is also often affected by seasonal food shortages. Abortion, stillbirths and miscarriages are more common among undernourished women than among those that are well fed. Hookworm and malaria may also weaken the mother and affect the child.

The pregnant adolescent

In the tropics many girls have their first baby when still very young and physically immature. This means that the diet must provide for the growth of the mother's own body as well as for that of the foetus. The earlier pregnancy occurs in the teens the more serious is the problem. If the mother is under fifteen years of age the newborn tends to be underweight and infant mortality is about twice that of the baby born to a mature mother.

Other complications for the pregnant adolescent are premature labour, toxaemia of pregnancy, iron-deficiency anaemia and prolonged labour. If repeated pregnancies occur during the adolescent years there is a further increased risk of iron-deficiency anaemia. The pregnant teenager should receive vitamin and mineral supplements as well as high intakes of energy and proteins to improve the chances of good health for herself and her baby.

Infant requirements

During lactation also the mother should be well fed, particularly if she is working. To satisfy the needs of an infant 5 kg in weight, she requires 800 ml of milk equivalent to 2.5 MJ per day. The composition of milk is very constant except that if the mother has too little thiamin, and vitamins A and C, these may be low in the milk. Particularly low thiamin levels can lead to disease in the infant (infantile beriberi). The composition of human milk is given in Table 4.6.

Breast feeding is extremely important and one or two small feeds can be given until the child is two years old, in addition to its other food.

So it is clear that repeated cycles of pregnancy and lactation are a heavy drain on the mother and the teaching of both contraceptive techniques and dietetics should have a high priority in all schools in the tropics.

Table 4.6 *Composition of human and cow's milk (per 100 ml)*

	Human	Cow's
Protein (g)	1.1	3.5
Fat (g)	4.0	3.5
Carbohydrate (g)	9.5	4.9
Calcium (mg)	33	118
Phosphorus (mg)	14	93
Iron (mg)	0.1	0.05
Vitamin A (μg)	50	30
Thiamin (mg)	0.01	0.03
Riboflavin (mg)	0.04	0.17
Niacin (mg)	0.2	0.1
Ascorbic acid (mg)	5	1

Breast feeding versus bottle feeding

Milks is the basic food of the infant. Although its natural food is mother's milk in many very advanced countries babies are fed with cow's milk from a bottle. Only recently has there been a trend by upper class families in these countries to revert to breast feeding. Table 4.6 gives the analysis of human and cow's milk.

These results show that cow's milk contains three times as much protein and twice as much carbohydrate, four times as much calcium and riboflavin and six times as much phosphorus as human milk. However, the latter contains five times the amount of ascorbic acid.

If the baby is fed with cow's milk the higher concentration of minerals and protein has two effects. First, it can lead to higher concentrations of phosphate in the blood with a risk of convulsions. Second the larger amounts of urea and minerals require excretion. This may overtax the kidneys. Therefore cow's milk must be diluted with water (one part of water to two parts of milk). Then to increase the energy content 3.5 g of sugar must be added to 100 ml of the diluted feed. Next the mixture must be boiled to sterilise it and finally it must be cooled to blood heat.

There are important reasons why infants should be breast fed wherever possible. Breast milk is nutritionally satisfactory, cheap, readily available and sterile and breast feeding leads to a good mother–child relationship. Immunity against certain diseases is passed from the mother to the suckling infant and breast feeding has also a definite contraceptive effect. While the mother is breast feeding ovulation is nearly always suppressed. It is not an absolutely foolproof method but in developing countries where contraception is difficult to practise but breast feeding carries on for a long time it significantly reduces pregnancy.

Many authorities believe that the most important argument against bottle feeding in the tropics is that fact that the great majority of women are not able to sterilise bottles properly and so cause diarrhoea in the infant. However there is evidence that after about 3 months breast milk as the sole food of the infant becomes inadequate. This gives rise to what has been called 'the weanling's dilemma' – the choice between being undernourished at the breast or infected from external sources of food. Traditional weaning food in a village can be as dangerous bacteriologically as commercial milk products made up with local water (see Chapter 17).

On the one hand it has been argued that it is better for the small child to be underfed than exposed to supplementary foods too early. On the other, it has been pointed out that satisfactory feeding with additional food helps the child to resist infection which indeed is not only caused by food but also by general unhygienic conditions.

When dehydration occurs during a bout of diarrhoea fluid can be given by intravenous injection or by mouth. For the latter WHO recommend 3.5 g sodium chloride, 2.5 g sodium bicarbonate, 1.5 g potassium chloride and 20 g glucose in 1 l of drinking water. In an emergency dissolve a handful of sugar (or about 30 g of rice powder) and a pinch of salt in 1 l of boiled water.

There is evidence to show that most women in the tropics produce satisfactory amounts of milk of good quality. However milk production tends to fall during the harvest and this suggests that pregnant and lactating women should be released from heavy agricultural work at that time. They should be well nourished and breast feed their infants as long as possible.

Feeding the elderly

As people get older they require less food. They tend to absorb their food quite satisfactorily but their teeth may have suffered and that may prevent them from taking an adequate diet. Apart from lower energy requirements their nutritional needs are little changed and their intake of protein, calcium, iron, folic acid and vitamin D particularly must be adequate.

Cultural and religious aspects

Practically all areas of the world have certain cultural or religious food avoidance.

Europeans do not eat cats and dogs although these were eaten in the famines following World War Two. In Africa fish is sometimes rejected because of its smell or snake-like appearance. Fish without scales may not be eaten.

In many parts of the tropics the blood of slaughtered animals is often wasted although it has a very high PER and is rich in several nutrients, particularly iron. On occasions goat meat and milk are not consumed and it is thought (falsely) that the white meat of birds is not as nutritious as the red meat of cattle. Often snails and highly nutritious insects are avoided.

Prejudices such as these may do no harm if alternative foods are freely available. It is utter folly to follow them if food is scarce. Two aspects are particularly serious. Firstly, children and nursing and pregnant mothers are primarily affected and these are particularly vulnerable groups. In West Africa pregnant women often avoid eggs and snails and in Malaysia watery fruits such as pumpkin, melon or cucumbers are not eaten by women who have given birth. On the Indian subcontinent in particular there are many food restrictions for pregnant or lactating women, and widows there also often reject animal products.

Secondly, although some secret societies, e.g. in Sierra Leone, do not eat groundnuts, millet and certain vegetables, avoidance of plant products is rare. It is mainly the important protein-rich animal food that is taboo.

Food avoidance for religious reasons is also common. Orthodox Hindus are vegetarian and strongly support the adoption of 'ahimsa' which means 'not killing'. Milk, yoghurt, butter or ghee are usually eaten because no killing is involved. Cheese is often unfamiliar or not eaten because rennet may be used in its manufacture. (Cheese can of course also be made without it.) Less-orthodox Hindus often take mutton, poultry, fish and occasionally pork, although beef is never eaten because the cow is regarded as a sacred animal.

Sikhs tend to have similar food avoidance but meat is taken more often.

Islamic custom prohibits pork and alcohol and like the Jewish religion insists on the ritual slaughter of animals.

In a vegetarian diet special attention must be paid to the protein content of the food. As wide a variety as possible of permitted foods of high protein content should be eaten. These include milk, yoghurt and cheese, as well as legumes, nuts and cereal products.

5

The texture, colour and flavour of food

People do not eat nutrients, they eat food and the food they eat must be pleasing to them. The texture, colour and flavour of the food must be satisfactory. Some characteristics of food are liked by everyone, some by certain groups of people and some are quite individual.

Take for instance the Nigerian people. Compared to British tastes, Nigerians prefer their meat more chewy and do not cook vegetables so much. In general fruit such as banana, mango, tangerine or orange should be coloured and not green. The colour suggests sweetness and juiciness. If tangerines are sold green, the seller often cuts one across to show that it is still juicy.

For no good reason, some Europeans like eggs with brown shells, or with dark yellow yolks and others with white shells or pale yolks. There is no such prejudice in Nigeria. But there some people like yellow gari or yam and others white gari. The Yoruba prefer their gari sour, the Ibos less so. Yoruba people on the whole eat more spicy foods than other Nigerians.

Some food preferences and dislikes are quite individual. A Nigerian friend of the writer dislikes smoked fish and the writer himself detests carrots.

So it is clear that different people have different tastes and the food scientist must consider them. First it is necessary to separate the characteristics of food into texture, colour and flavour and consider each separately.

In the laboratory the tasting of the food is often done in specially designed booths (Fig. 5.1). These are equipped with white or coloured light, fume extractor, and water. In this way each taster is isolated and less influenced by his colleagues or his surroundings. The tasters can be asked to state their likes or dislikes of the food often on a 5-point scale (1 = like extremely, 2 = like, 3 = neither like nor dislike, 4 = dislike,

Fig. 5.1. A typical tasting booth. These are often arranged in sets of four or six.

5 = dislike extremely). They can also be asked to distinguish food A from food B. They are then given three samples (AAB or ABB) and asked to separate them. Often statistical methods are needed if the result is not clearcut.

Texture

This has been defined as mechanical properties as perceived by our senses. It is often possible to see or smell whether a fruit is overripe, but an important test is to squeeze it. Experience will tell which type of

Fig. 5.2. The Instron tester. In this picture the instrument is set up for a penetration test. The middle horizontal bar in the right hand unit moves down and a vertical projection punctures the sample. The left hand unit records the result on a paper chart.

texture is satisfactory and which is not. Machines have been designed for the testing of texture by imitating fingers or teeth. An important instrument is the Instron (Fig. 5.2). This consists of two bars which can move towards or away from each other in a carefully controlled manner. Various test cells can be placed between these bars and the cells contain the food to be tested. The force exerted on the food in the cell can be measured and plotted against the deformation of the food. Figure 5.3 shows some of the test cells which can be used. The conditions of the test must be very accurately controlled for the results to be repeatable. The cell dimensions, rate of compression or extension, temperature as well as the shape of the test samples are carefully observed.

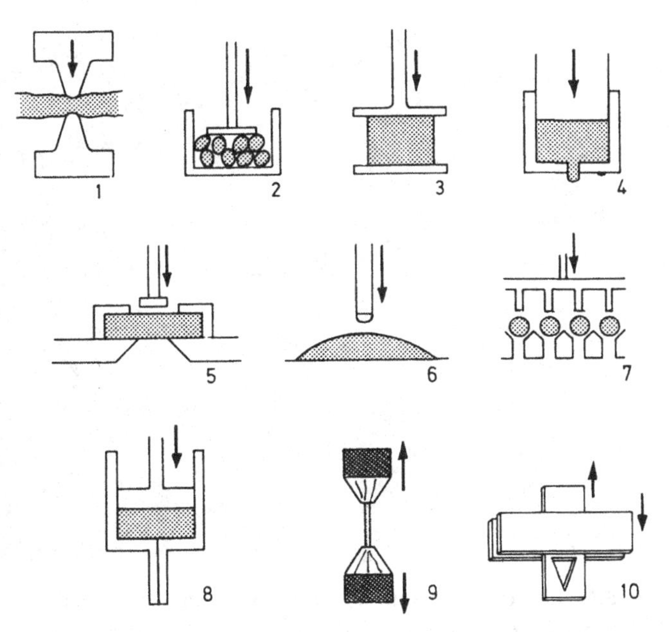

Fig. 5.3. Sample cells used with the Instron.
1. Meat tester. 2. Gap extruder (grapes). 3. Plate compressimeter. 4. Juice extractor (meat, etc.) 5. Shear strength tester. 6 Apple penetrometer. 7. Pea penetrometer. 8. Capillary extruder. 9. Extension tests. 10. Meat shear unit.

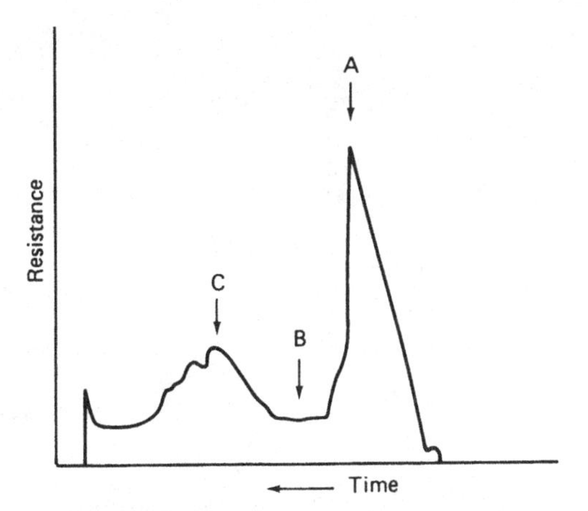

Fig. 5.4. Instron curve showing penetration of ripe tomato (see text).

Figure 5.4 shows a typical penetration curve on a ripe tomato. At first the penetrometer indents the tough skin of the fruit which breaks at point A. The resistance then drops as the plunger slides through the soft tomato pulp B. It then rises as the plunger meets the vascular bundles in the

Fig. 5.5. Photomicrograph of tomato skin. Bar = 100 μm. (Courtesy F.O. Flint.)

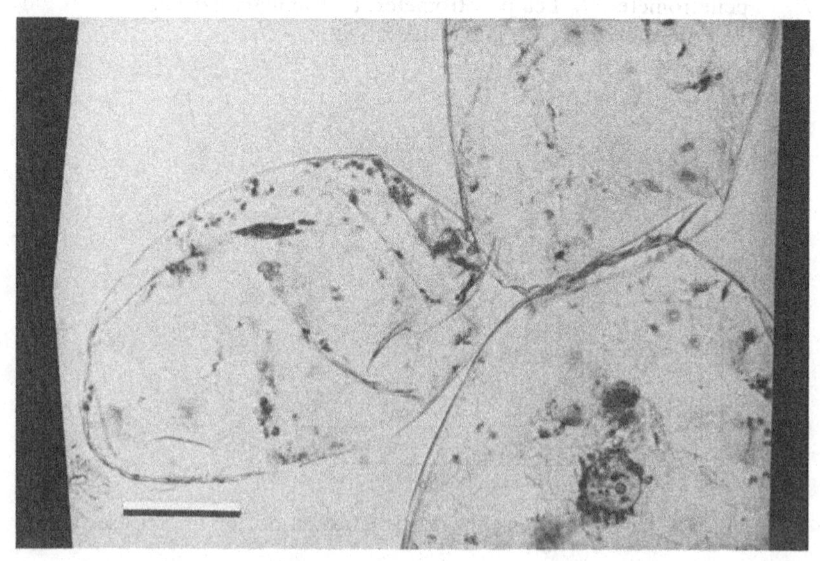

Fig. 5.6. Flesh cells from pulp of tomato. Bar = 100 μm. (Courtesy F. O. Flint.)

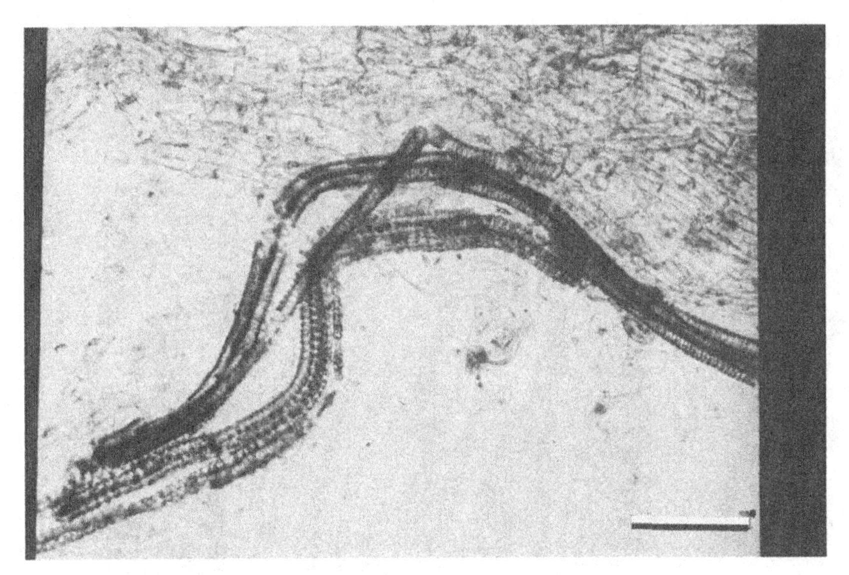

Fig. 5.7. Vascular bundles of tomato. Bar = 100 μm. (Courtesy F.O. Flint.)

centre of the fruit C. So mechanically speaking the tomato consists of three phases, skin, pulp and vascular bundles. These are shown in the photomicrographs Figs. 5.5, 5.6 and 5.7.

The most important factors in determining the texture of plant material are described below.

(1) Cell wall

When first laid down the cell wall consists mainly of pectin. No particular structure can be seen under the electron microscope. Then the primary wall is laid down which consists of cellulose fibrils. Young shoots are at this stage. When the cell has grown to its full volume, additional layers of cellulose as well as lignin are deposited. These are seen as well aligned layers under the electron microscope. Vascular tissue develops with extra solid thickening which makes older vegetables fibrous and stringy. There is therefore a relationship between the Instron test, consumer acceptability and the chemical determination of fibre (Fig. 2.5).

(2) Turgor pressure

The cell protoplast acts as a semi-permeable membrane and due to the osmotic pressure inside the plant cell water is absorbed. The aqueous

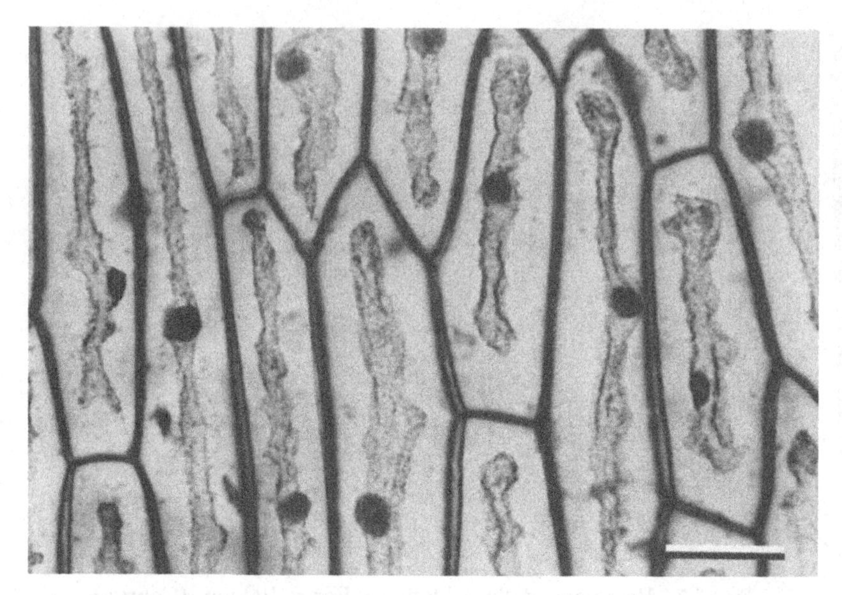

Fig. 5.8. Plasmolysed cells of onion skin (see text). Bar = 100 μm. (Courtesy F.O. Flint).

solution presses outwards against the resistance of the wall and this is called turgor pressure. Many leaves and fruits are kept rigid by turgor pressure. If a crisp leaf is immersed in boiling water, the cells are killed and the semi-permeability of the membrane is destroyed. As a result water leaks out of the cell, the turgor pressure falls and the fruit or leaf may become soft and flabby. Similarly if the tissue is immersed in strong sugar solution, water is drawn out of the cell. The cell is said to be plasmolysed. Figure 5.8 shows the plasmolysed cells of onion skin. The dark nucleus, the granular cytoplasm and the dense cell walls are clearly visible. For many plant tissues of this type, there is a good relation between the Instron test, consumer acceptance and moisture content.

(3) Cell inclusions

In yam, cassava and other tubers, firmness is due to the starch granules which are tightly packed inside the cell. On cooking, these gelatinise and the tissue softens.

(4) Intercellular adhesion

As the fruit ripens the middle lamella softens. (The middle lamella is the layer that sticks the walls of adjacent cells together.) Banana tissue has

Table 5.1 *Average composition of meat and blood (g per 100 g)*

Component	Lean meat	Pig's blood
Water	75	79
Protein	21	19
Carbohydrate	0.3	0.1
Lipid	3	0.5
Others	1.6	1.5

been well investigated. Here the softening is due to interconversion of the structural polysaccharides. The soluble pectins increase as the fruit ripens and softens.

Often several of the above factors are involved together in changes in texture. These may occur with natural ripening or on domestic or industrial processing. In fresh fruit the texture is often determined by the turgor pressure, but in cooked fruit or jam by the vascular bundles. In raw onions the texture is again due to turgor pressure, in pickled onions to the vascular bundles. Indeed on freezing or cooking, the cell walls are often ruptured and the importance of turgor pressure in the tissue decreases, while cell wall, cell inclusion and intercellular adhesion become more important.

With meat the problem is different. Table 5.1 shows the average composition of lean meat and pig's blood. This table shows that the texture of meat depends on its structure and not on its chemical composition.

Another problem with meat is rigor mortis, the stiffening of flesh shortly after death (see p. 180 et seq.). During rigor the elasticity of the meat changes but these changes are not reversed when rigor disappears on ageing of meat. It is thought that elasticity is related to stiffness but not to tenderness. The longer the muscle cells (sarcomeres) are and the less connective tissues the meat contains, the more tender it appears to be.

Another example of the relation between texture and structure is given by oils and fats. Oils show no structure under the microscope and their viscosity is due to molecular structure. The larger the molecules and the stronger their interaction the more viscous is the oil. Spreadable fats (butter, margarine) are said to be plastic. They do not flow under gravity but require a larger force for flow to occur. For any material to be plastic three conditions must apply:

(1) There must be a 2-phase system consisting of a continuous liquid phase and a 'solid-like' dispersed phase. This 'solid-like'

Table 5.2 *Examples of plastic foods with reference to their solid-like and liquid phases*

Example	Solid-like phase	Liquid phase
Butter	Fat and water globules with surface film	Oil
Lard	Fat crystals	Oil
Mayonnaise	Oil and surface film	Aqueous solution
Whipped cream	Air and surface film	Protein solution

phase need not be a true solid as long as it acts like one. For instance, a bubble of gas or liquid may act as a solid if surface tension makes it behave so.

(2) There must be a fine dispersion of solid in liquid. There must be no seepage or settling, and the whole mass must be held together by internal cohesion.

(3) The two phases must be in the correct proportion. If there is too much liquid the material will flow under gravity. If too much solid it is brittle.

Table 5.2 gives some examples of plastic foods with reference to their solid-like and liquid phases.

Colour

Because the colour of food is associated with the ripening of fruit or the deterioration of meat, fish, or vegetables, we powerfully associate colour with food quality.

Egg yolk and butter must be yellow, bread white and chin-chin brown. Green meat, blue bread and red margarine are not acceptable. So processed foods are often artificially coloured to improve their appearance. Some colours are obtained from natural sources. Annatto, a yellow colour, is obtained from the annatto tree, turmeric, also yellow, comes from the root of *Curcumena longa*, an East Indian herb. Cochineal, a red dye, is obtained from a South American insect *Coccus cacti*. Some natural dyes have been synthesised in the laboratory. They include β-carotene, riboflavin and curcumin. The great majority of dyes are artificial. They are more stable, more uniform and cheaper than natural or synthetic colours. Well over 100 have been used in the past but now their number has been greatly reduced. Possible health hazards made extensive testing necessary and that is very expensive.

There are basically four types of colour of major importance in food.

(1) Pyrrole derivatives

These are mainly the green pigment chlorophyll (chloros is green in Greek and phullon means leaf) found in plants and the red heme (haema means blood in Greek) which gives meat its colour. In nature both these are attached to a protein. The molecule of chlorophyll contains one atom of magnesium, that of heme one of iron. The structural formula of chlorophyll is given below.

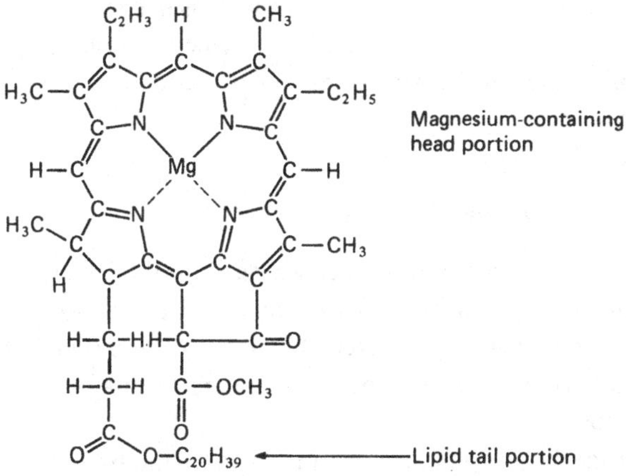

Magnesium-containing head portion

$O\text{-}C_{20}H_{39}$ ← Lipid tail portion

When processing green vegetables sodium carbonate is often added. This neutralises acids and helps retain the green colour. Texture and flavour may however be badly affected by this treatment. Blanching gives better colour retention. This is due to the destruction of colour-degrading enzymes or possibly to traces of copper. This is known to give an intense green colour with chlorophyll. Sometimes the colour of vegetables has been improved by blanching followed by canning in extremely dilute solutions of sodium or calcium hydroxide.

(2) Carotenoid pigments

This group of pigments is called after the main colour in carrots. It is widely distributed and found in all green leaves, most yellow and red fruits, yellow maize and yellow yam and many stems and roots. It is not synthesised by animals but may be absorbed by them. It is therefore found in crustaceans, egg yolk and many fish.

On processing it is frequently oxidised but the effect varies a great deal. For instance, much pigment is retained on drying apricots but not carrots.

On canning most is retained. On freezing green vegetables are blanched first to avoid carotenoid breakdown.

(3) Flavonoids and anthocyanins

These pigments comprise a large group of rather complicated organic molecules ('flavus' is yellow in Latin; 'anthos' is flower and 'cyan' is blue in Greek). In foods they may give rise to enzymatic browning and in the presence of iron catalysts to green or blue discolouration. They are the main colour in many red and blue berries, as well as in cherries and blood oranges. While heating often bleaches the pigments, low-temperature processing usually does not. Sometimes the pinking in canned pears is due to these pigments.

(4) Caramel and melanoidins

Caramel is obtained from heating sucrose as described in Chapter 2. Melanoids are brown pigments which are due to non-enzymatic browning reactions ('melas' means brown in Greek). There are very many of these. They are difficult to classify and the chemical reactions are often not properly understood.

A convenient method of isolating pigments is by the use of thin-layer chromatography (TLC). Ether extracts are spotted on a glass plate covered with an even layer of silica gel. If necessary the intensity is built up by repeated application, drying each time with a hair dryer (Fig. 5.8a). When a row of extracts and pure chemicals (controls) have been applied the plate is dipped into the solvent to develop it. Figure 5.8b shows after development, as controls β-carotene (near the solvent front), lycopene (a little lower and tailing) and canthaxanthin (which does not move in the solvent). The plant extracts show that paprica and carrot contain β-carotene, and canthaxanthin, tomato β-carotene and lycopene while tomato puree contains additionally a carotene breakdown product.

TLC has many uses and is often employed to demonstrate mycotoxins (see Chapter 6).

Flavour

The perception of flavour consists of four separate factors: taste, smell, touch and sound.

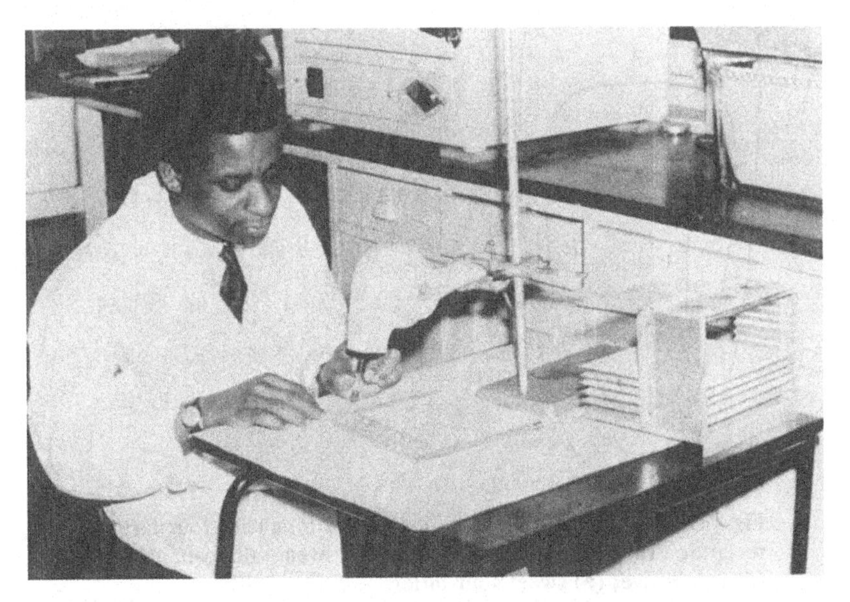

Fig. 5.8a. In thin layer chromatography the extracts are placed on the plate in successive drops while the hair drier evaporates the solvent. In this way the intensity of the spot is increased.

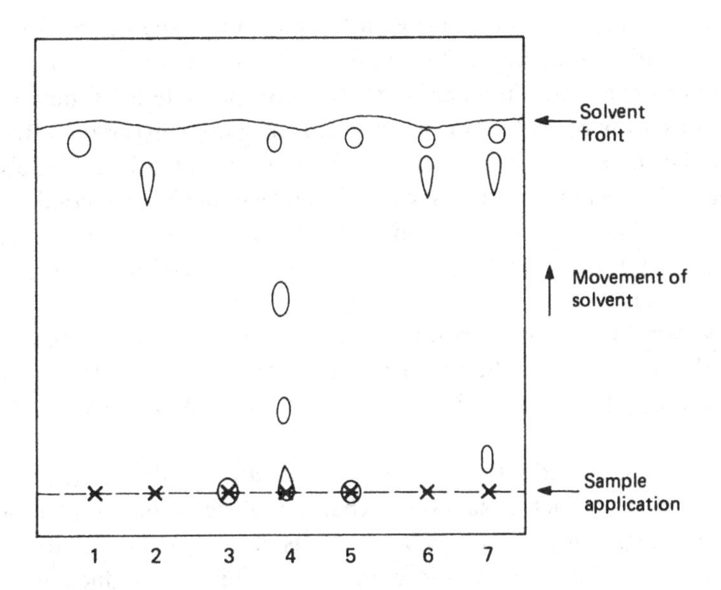

Fig. 5.8b. Thin-layer chromatogram. Solvent Petroleum spirit (40–60°C): Toluene 4 : 1 on silica gel 60G.
Controls: 1 = β-carotene, 2 = lycopene, 3 = canthaxanthin.
Solvent extracts: 4 = paprika, 5 = carrot, 6 = tomato, 7 = tomato puree (see text).

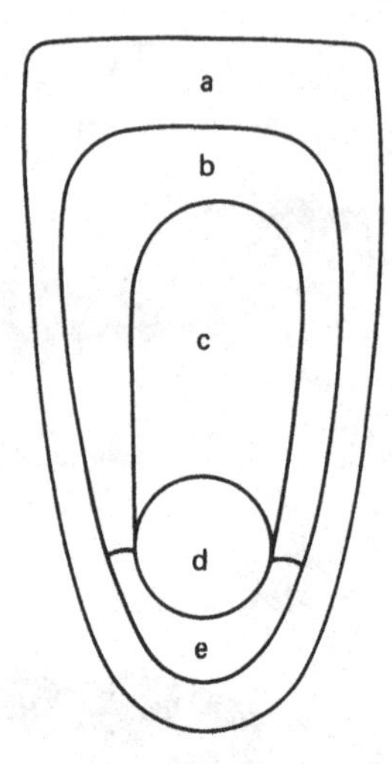

Fig. 5.9. The approximate location on the tongue of four taste receptors: (a) bitter sweet salty sour; (b) sweet salty sour; (c) sour; (d) insensitive; (e) sweet sour bitter.

Taste

This, sometimes called 'true taste' or 'tongue taste' is perceived through the taste buds of the tongue. Normally, the stimulant must be an aqueous solution to stimulate the taste buds. It is also possible to stimulate the taste buds 'from below'. For instance, if 10 mg of thiamin (vitamin B_1) are injected intravenously it is possible to perceive a peanut-like taste about 10 s later in the mouth. In this way one can taste one's own blood!

The four basic tastes perceived by the tongue are sweet, bitter, salty and sour. Figure 5.9 shows the approximate location of the receptors for these four tastes.

However the taste buds are not confined to the tongue but are scattered around the oral cavity. The primary receptors for sour and bitter are near the junction of the soft and hard palate at the back of the roof of the mouth.

There are many stimuli for sweetness that are mentioned in Chapter 10. Most of these such as sugars, saccharin or glycyrrhizin are non-ionic. Sweet ionic stimuli are few, but include salts of lead and beryllium, some very poisonous. Indeed the old name for beryllium was glucinum, the sweet element.

For bitterness the non-ionic stimuli include caffeine, quinine and strychnine (a deadly poison) as well as the various tannins. Bitter-tasting ions include silver, ferric iron and iodine. Most bitter-tasting substances are only slightly soluble in water and are perceived only slowly in the mouth. However, once noticed, the sensation lasts for quite a long time.

The threshold amount of quinine hydrochloride, i.e., that amount which can just be tasted, is about 0.016 mg and for caffeine about 0.040 mg. That compares with the threshold for sucrose of about 7 mg. The threshold for bitter substances varies a great deal from person to person. For instance, a few people detect a bitter after-taste with saccharin-sweetened foods or drinks but most do not.

Salty taste is always due to ions. The most important is sodium chloride; others, some very poisonous, include cations of fluoride, potassium, magnesium and ammonium. Salty ions include carbonates, nitrate and sulphate. The threshold for sodium chloride is about 2 mg.

Smell

This is detected in the nasal cavity. The sensitive area is quite small, about 25 mm square, the size of a postage stamp. It lies at the very top of the nasal cavity, just below the eyes. The molecules causing the smell can reach this area either through the nostrils (and a good sniff will give a good response) or past the back of the oral cavity upwards. This is less dramatic but important during swallowing. The relation between molecular structure and olfactory response is not known. Therefore the activity of the manufacturer of essences or of the perfumier is an art rather than a science. It is possible to detect incredibly dilute concentrations of smells. Vanillin, artificial musk and some sulphur-containing compounds can be detected at concentrations of one five-hundred millionth of a milligram (2×10^{-12} g) in 50 ml of air. Even that is little compared with the sensitivity of some insects which can smell their mate over very great distances.

A great deal of work has been done on odours. Particularly gas-liquid chromatography (GLC) has been used to isolate the various odour constituents in spices. Other scientists have prepared sets of pure chemicals to provide odour standards. Finally, the words used for odours have been listed and compared. This is difficult because people may not describe the same odour with precisely the same words.

Spices are plant organs which are added in very small quantities to improve the flavour and digestibility of food. Their effect is mainly due to fairly complicated organic chemicals which may have a particularly pleasant or pungent taste or smell. Since these chemicals are volatile and the spices themselves are easily attacked by insects, the spice should be kept in tightly closed containers. Particularly since the advent of GLC

Table 5.3 *Some important spices*

Spices	Botanical name	Plant part	Most important constituent
Ginger	*Zingiber officinale*	Rhizome	Gingerol
Pepper (black and white)	*Piper nigrum*	Fruit	Piperine
Sweet and chili pepper	*Capsicum* spp.	Fruit	Capsaicin
Cloves	*Eugenia caryophyllus*	Dried buds	Eugenol
Aniseed	*Pimpinella anisum*	Fruit	*trans*-Anethole
Cinnamon	*Cinnamomum zeylanicum*	Bark	Cinnamyl aldehyde
Caraway	*Carum carvi*	Fruit	(+)-carvone

many of the flavour components have been isolated. There may be up to 200 separate flavour constituents in one spice, but sometimes one is predominant (Table 5.3).

Touch is involved in flavour perception as regards both temperature and texture. Texture has already been considered. It is related to the rate at which food is broken down in the mouth and the speed with which the flavour in the food reaches the sensors of the tongue or nose. Furthermore, grittiness, lumpiness, and stickiness are important and so is the temperature of a food. A very hot drink may stimulate the pain receptors to such an extent that no flavour at all is perceived. Hot sour dishes, e.g. lemon tea, do not taste a lot until cooled down in the mouth. Sourness receptors do not react much above body temperature. Very cold foods may have very little flavour unless they are allowed to reach a higher temperature in the mouth.

Salty, bitter and sweet tastes are also greatly affected by *temperature*, so tasting tests should allow for this. Food should be held in the mouth until it has reached body temperature.

Sound also has a bearing on flavour perception. The sound vibrations produced when peanuts, cornflakes, meat and other foodstuffs are chewed have been recorded and analysed for amplitude at various frequencies. There are differences between foods and between eaters. This type of analysis gives information on the course of chewing both within one single bite and within a series of bites.

6

Food microbiology

Microorganisms are very small, living organisms which can only be seen with the aid of a microscope. Although they are usually regarded as plants, they do not contain chlorophyll. They are either saprophytes or parasites. A few of them exhibit locomotion. Many microorganisms are concerned with food spoilage and give rise to sour taste, sediment, discoloration, cloudiness, slime or scum. Some cause useful fermentations (Chapter 14). In this way they produce desirable chemicals or remove undesirable ones. Microorganisms are conveniently divided into three groups: moulds, yeasts and bacteria. The first two are classified as fungi.

Moulds

The mould body is called a mycelium. It consists of fine threads called hyphae which may have cross-walls (septate) or be devoid of them (non-septate). The non-septate hyphae contain many nuclei. Reproduction may occasionally occur by transplantation of pieces of mycelium, but by far the most important is by way of the vast number of spores. These are in function similar to plant seeds, but are usually (but by no means always) produced asexually.

There are three main types of asexual spores:

(1) Sporangiospores. These are produced in a sac on top of a fertile hypha called a sporangium (example: *Phycomyces*).

(2) Conidia. These are not enclosed but budded off from the top of conidiophores (example: *Aspergillus*).

(3) Arthrospores or oidia. These are produced by fragmentation of a hypha (example: *geotrichum*).

Some fungi also produce resting spores, often when conditions become

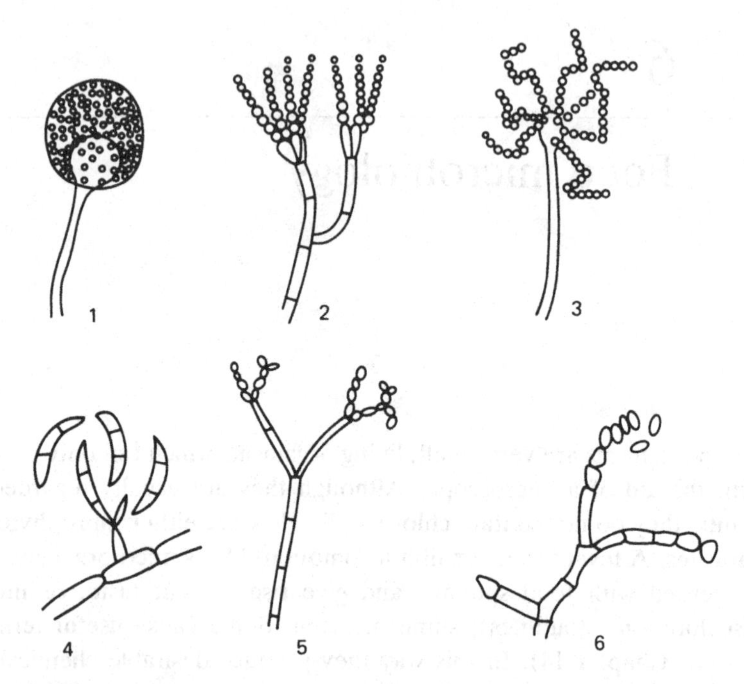

Fig. 6.1. Asexual reproductive organs of some common fungi (not to scale):
1, *Phycomyces*; 2, *Penicillium*; 3, *Aspergillus*; 4, *Fusarium*; 5, *Cladosporium*; 6, *Geotrichum*.

adverse. These spores, called clamydospores, are very resistant and can germinate when conditions become more favourable. Fungi tend to grow best at 25–30°C. These are referred to as mesophilic ('meso' is Greek for middle, and 'philos' for to love). Some are psychrophilic ('psuchros' is Greek for cold). These grow best at about +20° but will still grow, if slowly, at −10°C. Some are thermophilic ('thermos' is hot in Greek) and have a high optimum temperature. Most fungi are killed by moist heat at 100°C. Figures 6.1 and 6.2 show some of the asexual reproductive organs of fungi.

Yeasts

These are fungi which are usually one-celled (unicellular) and reproduce by budding and subsequent splitting (fission). Sometimes, when conditons are right some mycelial fungi can bud like yeast (see *Geotrichum*, Fig. 6.1) or yeast can give rise to a mycelium.

Figure 6.3 shows a freeze-etched preparation of baker's yeast (*Sac-*

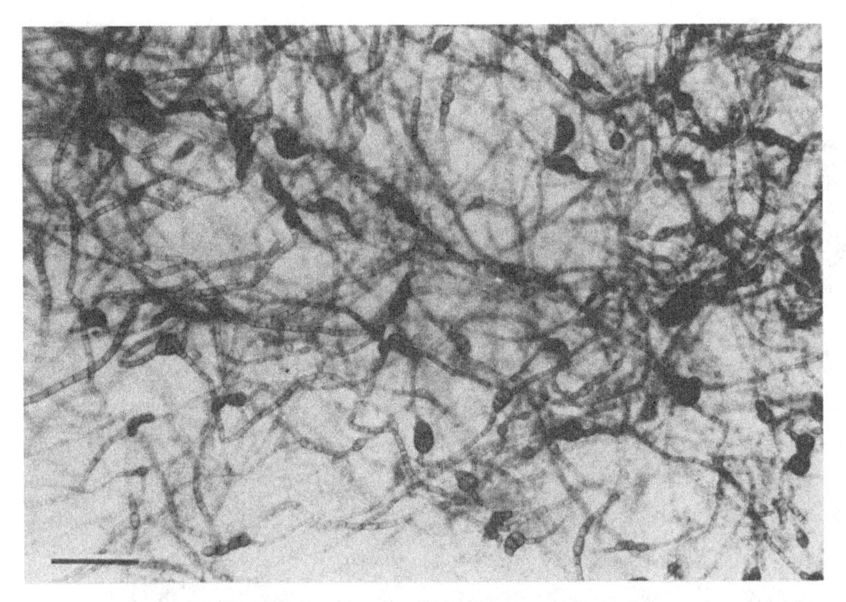

Fig. 6.2. Photomicrograph of the fungus *Alterania* growing on agar. The preparation has been stained with lactophenol. The spores and septate hyphae are clearly visible. Bar = 100 μm.

charomyces cerevisiae) under the electron mircoscope. At the top is the mother cell with its large vacuole and various cytoplasmic inclusions. Below is the smaller daugher cell being budded off.

Most yeasts are about 7 μm in diameter and oval in shape, although some are pear-shaped, lemon-shaped, cylindrical or even triangular. They may grow aerobically as a film or scum on top of a liquid or anaerobically within it. They tend to be inhibited above 35°C and all are killed by moist heat at 85°C. They tolerate a greater degree of acidity and a higher concentration of sugar than most other fungi or bacteria. That is why they sometimes grow on jam.

Most yeasts of industrial importance belong to the same genus as baker's yeast (*Saccharomyces*). The term 'wild yeast' is used for any extraneous yeast which contaminates the pure culture.

Bacteria

These organisms are much smaller than yeast or moulds usually being between 1 and 3 μm in size. Figure 6.4 illustrates the range of shape of some bacteria and Figure 6.5 shows an electron micrograph of freshly made yoghurt. The fat globule from the milk is apparent at the top left

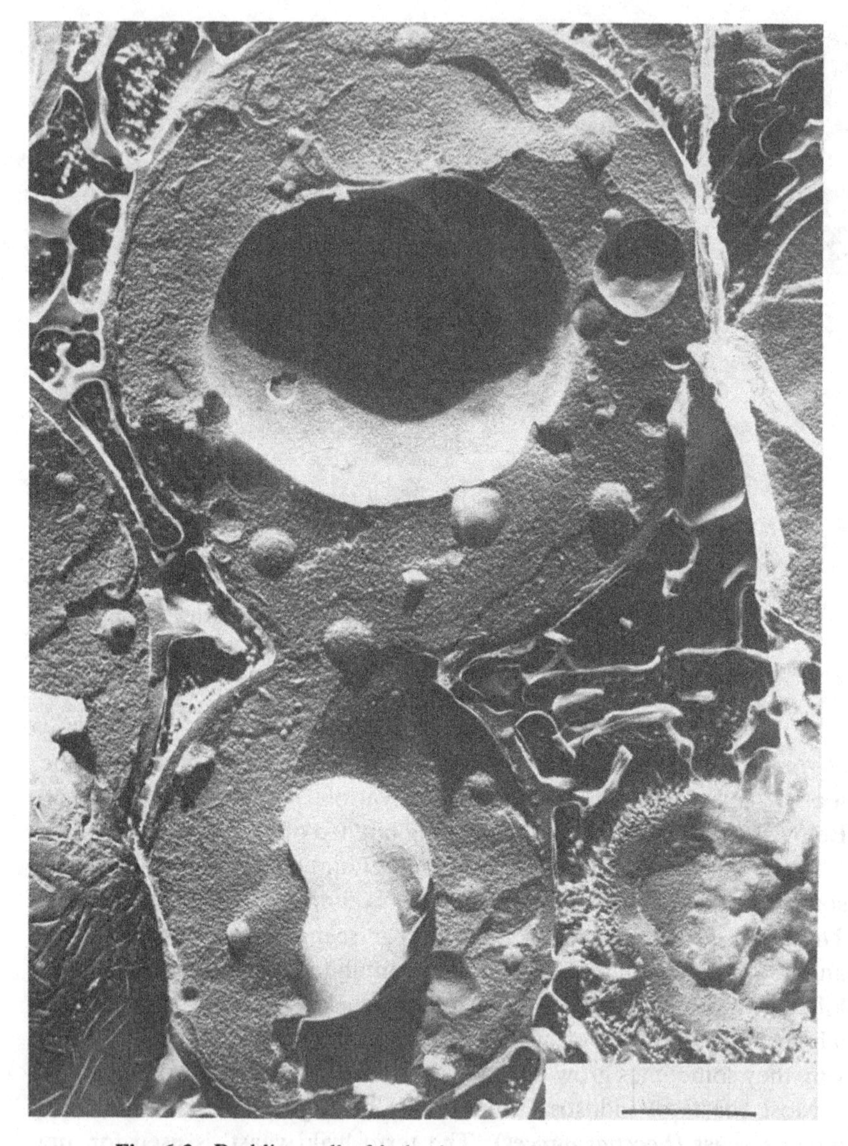

Fig. 6.3. Budding cell of baker's yeast. Mother cell at the top showing central vacuole. Daughter cell at the bottom. Bar = 1 μm. (Courtesy R. Reed.)

and two cells of *Lactobacillus* on the right. The granular background is casein particles from the milk.

Many bacteria are useful in producing waste products such as acids which sour foods such as soft cheeses, cream, ogi or kenkey (Chapter 14). Some produce extremely dangerous poisons (see below). Much spoilage

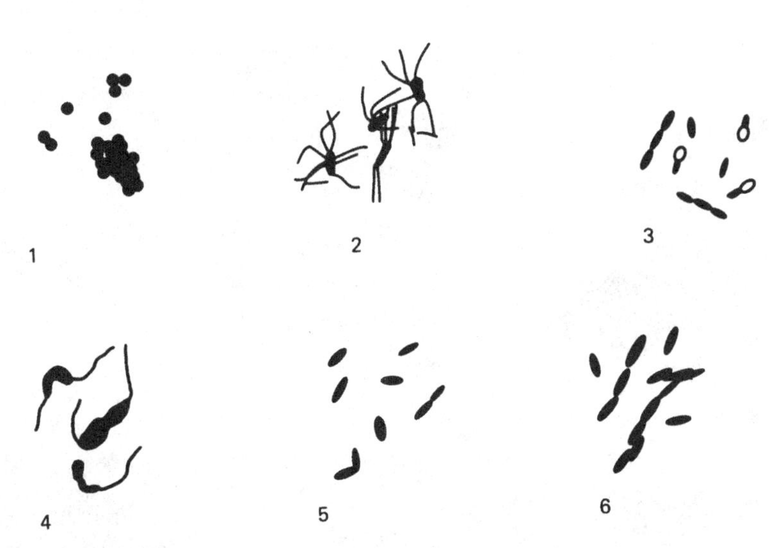

Fig. 6.4. Bacteria (not to scale): 1, *Staphylococcus aureus*; 2, *Proteus vulgaris* (showing flagella); 3, *Clostridium tetani* (tetanus, showing resting spores); 4, *Vibrio cholerae* (or comma bacillus, the cause of cholera); 5, *Pseudomonas fluorescens*; 6, *Lactobacillus bulgaricus*.

in canned foods is caused by spore-forming bacteria due to the heat resistance of their spores.

When conditions are favourable, bacteria can divide once every 20 min. *Clostridium perfringens* at its optimum temperature of 44–45°C divides every 12–13 min. Therefore a small number of bacteria can give rise very quickly to a very large population (one cell dividing every 30 min would produce in 24 h 280 000 000 000 cells!). Some bacteria such as species of *Clostridium* and *Bacillus* have very resistant spores which help them survive in very adverse conditions (see Food preservation, Chapter 15).

In order to count the bacteria in a given food a representative sample must first be taken. This is relatively easy with liquids where sterile pipettes are used. With solid foods some of the methods used involve sterile cork borers, swabs, stomacher bags or attrition with sterile sand. The food is then blended with a sterile diluent. Usually the volume of diluent used is nine times the weight of sample (e.g. 25 g of food in 225 ml of diluent giving 10^{-1} homogenate). From this further dilutions are prepared (10^{-2}, 10^{-3}, 10^{-4} etc.). Each of the dilutions is then allowed to grow at constant temperature on agar plates and the number of bacteria in the original sample can be ascertained. This is the commonest method of counting bacteria (1) (Fig. 6.6). Other methods are as follows:

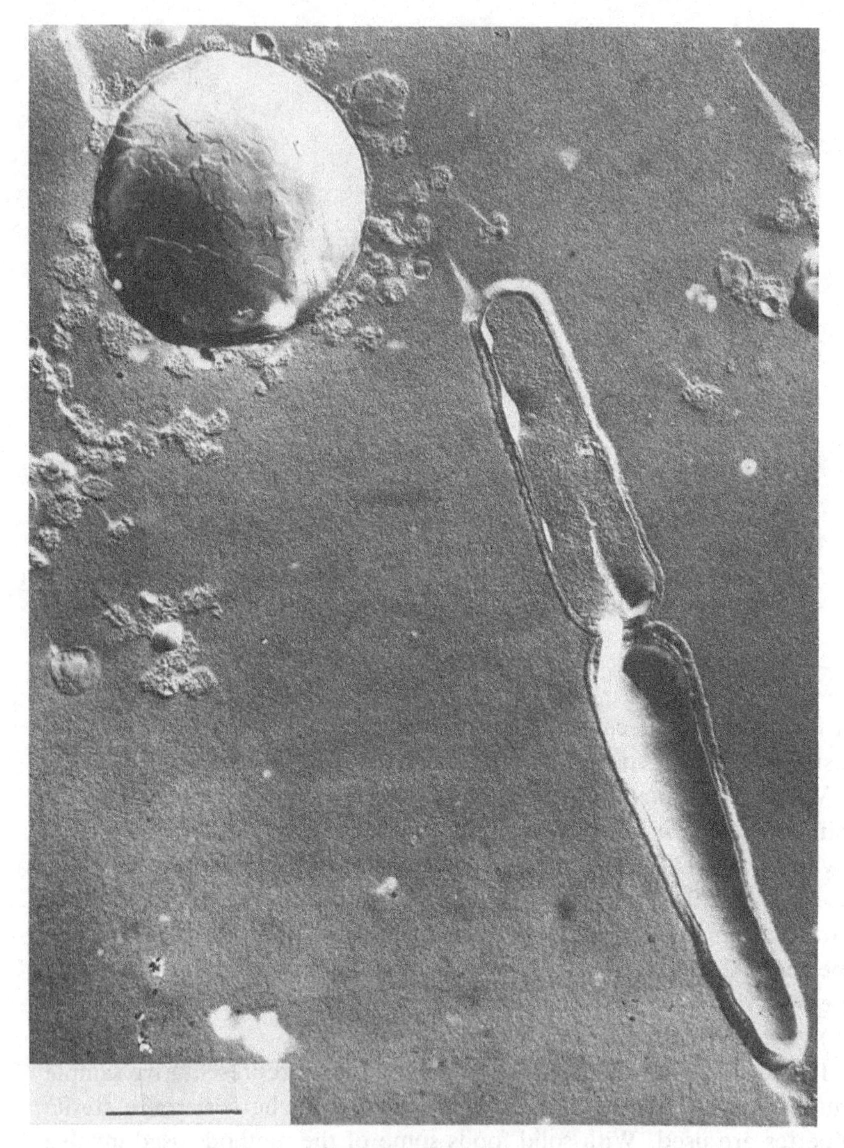

Fig. 6.5. Electron micrograph of yoghurt showing fat globule (top left) and cells of *Lactobacillus* from the starter culture. Bar = 1 μm. (Courtesy R. Reed.)

(2) ATP photometry. This works on the principle that all living organisms contain ATP (adenosine triphosphate, a compound concerned with energy transfer). When this comes into contact with a luciferin-luciferase enzyme substrate (isolated

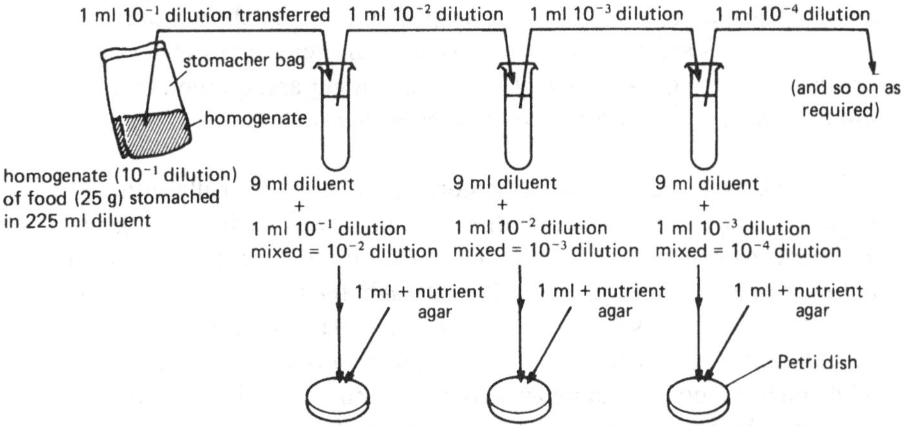

Fig. 6.6. Schematic representation of the 'pourplate' method used in the enumeration of bacteria. (From: Hayes, P. R. *Food Microbiology and Hygiene*. 1985. Elsevier Applied Science Publishers Ltd)

from glowworms) light is emitted. The light is proportional to the ATP and can be measured.

(3) Impedance measurement. This is based on the fact that living organisms alter the chemical composition of the growth medium. This causes a change in the electrical properties of this medium which is measured as impedance.

(4) Direct epifluorescent filter technique (DEFT). The food suspension is filtered and the filtrate containing the bacteria is passed through a fine membrane filter. The bacteria are retained on the membrane, stained with a fluorescent dye and counted using a special (epifluorescence) microscope.

(5) Microcalorimetry. Here the minute amounts of heat produced by growing bacteria are measured.

(6) Radiometry. With this method radioactive carbon dioxide produced by bacteria from a radioactively labelled carbon source incorporated in the medium is measured.

Methods (2)–(5) although much quicker than the plate count are based on high technology, are expensive and require a considerable scientific infrastructure. Methods (5) and (6) are little used.

Indicator organisms Since identification of bacteria is difficult and often beyond the scope of the usual bacteriological laboratory, foods are often examined for bacteria the presence of which indicates the possibility of food poisoning. These bacteria are referred to as indicator organisms. They grow normally in the human gut and if found on food indicate faecal contamination (e.g. *Escherichia coli*).

Oxygen requirements Microorganisms have typically different requirements for air or rather the oxygen contained in it. Obligate aerobes can only live in air or oxygen while facultative anaerobes thrive in either absence or presence of air although they thrive better in its absence. Both groups are represented by bacteria and yeasts rather than moulds. Obligate anaerobes thrive only in the absence of air or oxygen while microaerophils require an oxygen tension less than that of air.

A knowledge of the heat resistance of bacteria is most important in heat sterilisation. Psychrophiles, although having an optimum growth at very roughly 15–35°C. Many mesophiles grow between 15 and 50°C with an optimum at 30–45°C. Many thermophiles have an optimum temperature of about 55°C. Some have heat-resistant spores and their control in canning is all-important. The heat resistance of spores is affected by their age, the composition of the medium, pH, preservatives such as sulphur dioxide, some spices with preservative properties (e.g. mustard, cinnamon, garlic), or antibiotics.

Some microorganisms produce toxins causing pathogenic effects whilst others cause spoilage of canned foods mainly by acid production.

Bacterial food poisoning and food-borne infections

Bacterial food poisoning is normally divided into two major groups:

(1) The infectious type where the bacteria grow in the food and continue to grow in the host. Examples are infections with *Salmonella*, *Vibrio parahaemolyticus* and *Clostridium perfringens*.

(2) The toxin type. Here the bacteria grow in the food and produce toxin but live bacteria do not necessarily need to be ingested to cause the illness. Examples are *Clostridium botulinum* and *Staphylococcus aureus*.

In food-borne infections the food acts as a carrier and no replication is necessary. Examples are *Shigella*, typhoid and cholera.

Bacterial food poisonings can occur anywhere in the world but are predominant in hot climates where microorganisms grow rapidly. Initially the alimentary canal is affected and typical early symptoms are nausea, vomiting and diarrhoea. The incubation time before the onset of clinical symptoms varies from one hour to 3–4 days. Recovery usually takes place after 1–3 days except with botulism where recovery may take as long as ten days – if at all. In over half of cases botulism ends in death through respiratory failure. With salmonellosis mortality is below 1%. With other food infections or food poisoning mortality is fortunately very low.

The most important organisms are discussed below.

(1) Food infections

Salmonella exists as over 2000 different strains named after the place where the strain was first isolated (e.g. *S. newport*, *S. dublin*). The organism is a rod-shaped non-spore former and usually produces gas. Optimum growth temperature, as with most food-poisoning organisms, is 38°C. They are relatively sensitive to heat and are killed at 60°C in 15–20 min. They do not grow below 5°C. Being found in human carriers, pets and farm animals, particularly poultry, *Salmonella* multiplies in both man and his food.

Vibrio parahaemolyticus, a slender curved rod, is a facultative anaerobe which does not form spores. Temperature range of growth is 8–44°C with an optimum at 37°C. The organism is readily killed at temperatures above 50°C. Because it grows best in the presence of 3–6% salt it is found mainly in warm coastal waters. It can contaminate fish and shellfish.

Clostridium perfringens (syn. *Cl. welchii*) is an obligate anaerobe which forms spores. Food poisoning from this organism is caused by eating food containing at least 10^7–10^8 viable cells. There are five strains and the heat resistance of the spores varies greatly. Many require 1–5 h at 100°C for their destruction. Most commonly the cause of poisoning is meat which has been cooked and then kept warm for too long before consumption. The organism has also been shown to occur in soil, faeces and on birds and rats. Mortality is very low.

(2) Toxin producers

Clostridium botulinum is a spore-forming rod, an obligate anaerobe. There are seven strains lettered A to G, on the basis of the toxin they produce. Type E can still grow at 3.3°C while the optimum of most types is from 30 to 37°C at near neutral pH. The toxin produced is probably the most lethal known to man: as little as one microgram can kill. The organism as well as its toxin are destroyed by boiling.

Staphylococcus aureus. Man is the main source of this organism. It is frequently found in the nasal passage of carriers and in boils and wounds. Therefore the organism can be found in any food but most often in cooked or cured meat. Poisoning takes place only if there are at least 10^7–10^8 organisms per g of food. That means that the food must be a good growth medium and must have been standing about for some time. The toxin is called enterotoxin because it causes inflammation of the lining of the gut. It is very stable to heat and will withstand boiling for 20 to 60 m. Mortality is low.

(3) Food-borne infections

Shigella causes bacillary dysentery. This group of organisms has been known since the 4th century B.C. and whole armies have been laid low by them. The organisms are non-motile and do not form spores. They are often spread by contaminated water, milk and salads. Poor personal hygiene is a contributory factor.

Salmonella typhi is the cause of typhoid fever. It is spread by sewage-contaminated water, milk and ice-cream. Chlorination of water and pasteurisation of milk have greatly reduced outbreaks in developed countries. The organism may be excreted by recovered patients and carriers who themselves show no symptoms. Once a food product is handled and contaminated by them, the organism can be spread to other foods.

Vibrio cholerae or the 'comma organism' is a short curved motile bacterium. It is transmitted mainly through drinking water and occasionally food. A recent form has become widespread. It originated in Indonesia from where it spread via India, the Middle East and Africa to Western Europe. Inoculation is effective but lasts only about six months.

In West Africa one occasionally hears the view that 'African germs' are not dangerous to Africans. One is told that an African villager can drink riverwater without pretreatment while the white man must boil it first. So why hygiene? It is true that all of us have had a number of local diseases and against some of these we are later immune. We have either developed antibodies in our blood or have successfully competing microorganisms in our gut. However small children have not yet obtained immunity and hence the horrendously high infantile mortality in many developing countries. Furthermore none of us is sufficiently immune and food poisoning is as dangerous to Africans or Indians as to anyone else. Disease is caused by germs and not by evil spirits.

Fig. 6.7. The fungus *Aspergillus flavus* showing conidiophores. This microorganism produces aflatoxin. Bar = 100 μm.

Mycotoxins

Mycotoxins are fungal metabolites which are toxic to animals including man. More than 150 species of fungus have been shown to produce toxins although some only affect animals but not man and some have only been shown to be toxic in laboratory experiments, but have not caused a known disease. Since most fungi prefer warm temperatures and high humidity, the wet tropics are particularly liable to outbreaks of mycotoxicoses (i.e. diseases caused by mycotoxins). These may cause damage to liver, kidney, intestine and brain of both animals and man. Some cause eczema, some are carcinogenic (cancer-producing) and some are teratogenic (i.e. they produce abnormal births; 'teras' is Greek for monster).

Since one cannot conduct dangerous experiments with humans the evidence that, for example, aflatoxin a mycotoxin produced by *Aspergillus flavus* (Fig. 6.7) and *A. parasiticus*, causes liver cancer is indirect (Table 6.1).

The fungi most often responsible for mycotoxicoses are species of *Aspergillus*, *Fusarium* and *Penicillium*. They or their toxin can contaminate cereals, legumes, tubers, dairy products, meat and indeed most foods. The best control of this problem is prevention of circumstances in which the toxin can be produced. For instance grain must be harvested

Table 6.1 *Estimated aflatoxin ingestion (ng/kg body weight/day) and incidence of liver cancer (10^5 of population/year)*

Area	Aflatoxin intake	Liver cancer
Kenya (high altitude)*	3.5	1.2
Thailand (Songkhla)	5.0	2.0
Swaziland (High Veld)	5.1	2.2
Kenya (middle altitude)	5.9	2.5
Swaziland (Mid Veld)	8.9	3.8
Kenya (low altitude)	10.0	4.0
Swaziland (Lebombo)	15.4	4.3
Thailand (Ratburi)	45.0	6.0
Swaziland (Low Veld)	43.1	9.2
Mozambique	222.4	13.0

**Note:* High-altitude areas have a cooler climate than low-lying regions (see Chapter 1).

when ripe and must not be allowed to remain in the field. Both grain and legumes must be stored dry. Food affected by fungi must not be eaten. The infected part must be removed. Often the white or coloured mycelium can be seen with the naked eye on cereal grains, or groundnuts are misshapen or show signs of discolouration.

If prevention fails one must try to remove the toxin. Physical removal is one method. Another is extraction with organic solvents (chloroform, ethyl acetate, methanol with small amounts of dilute acid). Treatment with chlorine, ammonia or hypochlorite can be used. Cooking, sterilisation, parboiling, sun-drying and grain milling all reduce toxin levels. A combination of these methods can be used, for example with rice. The grain can be hand cleaned (i.e. contaminated seeds removed), parboiled, sun dried and milled to remove the husk. (See also Appendix 3.)

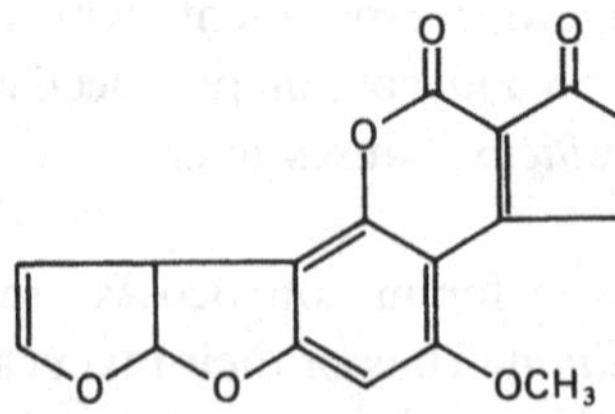

Aflatoxin B_1. The other major aflatoxins differ in the number and position of the side groups.

PART TWO

The composition of food

7

Cereals and legumes

Cereals

Cereal grains are the seeds of grasses large enough to be handled economically. They are the most important staple food and all major civilisations have been based on cereals. The word is derived from the Latin 'Cerealia munera' meaning the gift of the goddess Ceres.

Although constructed on much the same plan anatomically, the cereal grains differ considerably in size and shape. The weight in grams of 1000 seeds is approximately 0.3 g for t'ef, 10 g for millets, 20–40 g for rice, barley, wheat, rye, sorghum and oats and as high as 300 g for maize. The habitat of the cereals also varies considerably. While sorghum and the millets grow under almost desert conditions, wheat and barley are found in more temperate climates. Maize prefers it relatively hot and moist, while lowland rice is a swamp plant, growing in shallow water (Fig. 7.1).

Composition of the seeds varies as much within each cereal species as it does from one species to another (Figs 7.2, 7.3). On the whole, moisture content varies from 10 to 15%, protein from 8 to 15%, fat from 3 to 5%, although oats and maize may have a fat content of 8 or 9%. Carbohydrate, mainly starch, is of the order of 70%, and both fibre and ash 2%. The protein is of good quality although lysine is deficient and methionine, threonine and tryptophan may also be low. While vitamins A, C and D are either absent or present only in trace amounts, the cereals are a major source of some B vitamins (B_1, B_2 and niacin). There is no vitamin C but this is produced very rapidly during germination. Minerals particularly calcium and iron, are low although the millets contain useful amounts. The effect of milling can affect nutritive properties to a considerable extent (Chapter 13). Antinutrients in cereals may be phytic acid and tannins(see below).

The composition of cereal products as eaten depends to a large extent

Fig. 7.1. Cereals are often depicted on coins. Left to right, top: Mali – Millet, Israel – Wheat; bottom: Japan – Rice, Swaziland – Maize.

on the extraction of the flour (i.e. its bran content) and the amount of water and fat present. Biscuits have a water content of 3% and cereal porridge one of 60%. If the product is fried or deep-fried the fat content and therefore the energy value may be very high (see Table 7.1).

There is great activity in breeding new cereal varieties with specifically desirable properties. For instance, maize may be obtained with a protein content varying from 5 to 25% and 'high fat' maize may have a fat content of up to 17%. There are also high lysine, high amylose and high amylopectin maize varieties. Much the same range of composition applies to several of the other cereal grains.

Phytic acid Much of the phosphorus present in cereals is in the form of phytic acid. This must first be hydrolised in the body to release the phosphorus and it has been suggested that bacteria present in the intestinal tract fulfil that role. A rich source of phytic acid is the bran of cereals. Most authorities believe that phytic acid binds calcium, iron and zinc and makes them unavailable in human nutrition. It is probable that phytic acid binds these elements but that the body adjusts itself. With wheat flour the enzyme phytase present in the wheat destroys phytic acid during bread fermentation. Phytase is not present in maize unless the grains have sprouted.

Fig. 7.2. Ears of grain: A, Bullrush millet; B, wheat; C, rye; D, barley; E, rice; F, sorghum; G, oats. (Note: The black structures in rye (C) are due to a parasitic fungus, *Claviceps* which produces mycotoxins.)

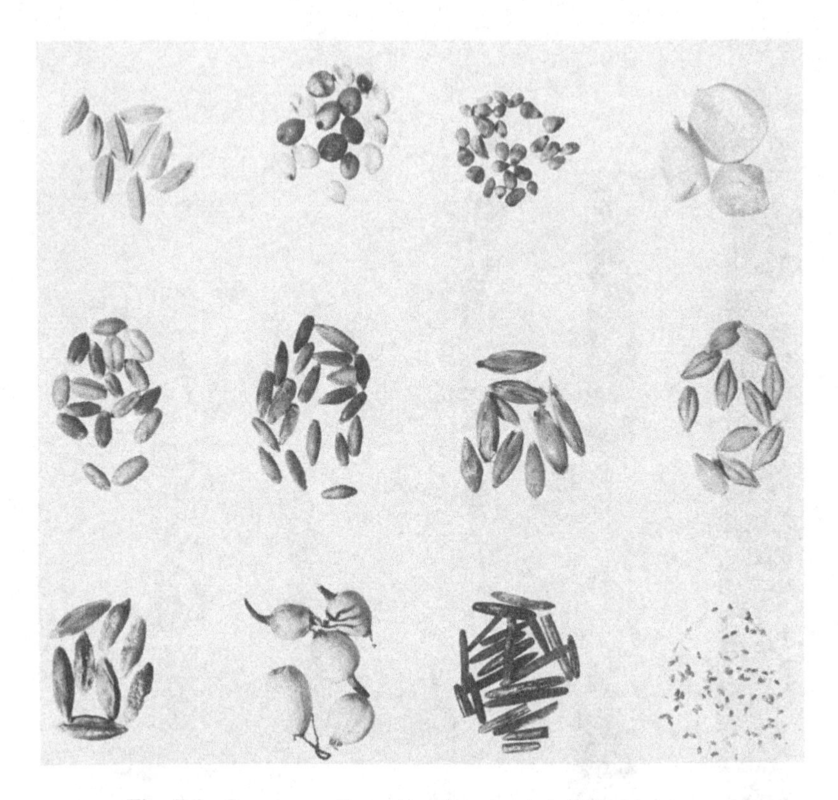

Fig. 7.3. Some cereal seeds. Top row left to right: rice, sorghum, millet, maize. Middle: wheat, rye, oats, barley. Bottom: Triticale, Job's tears, wild rice, t'ef.

Since the phytic acid is present mainly in the outer layers of the grain, much of it is removed in milling. Table 7.2 gives the amount of phosphorus (mg/100 g) and the percentage as phytic acid phosphorus in various plant materials. Phytic acid is not present in animal products.

Legumes and nuts tend to contain very variable quantities of both phosphorus and phytic acid phosphorus. Soya flour is particularly high in phosphorus (600–650 mg/100 g) and about 30% is in the form of phytic acid phosphorus. Most fruits and green vegetables are low in phosphorus and contain no phytic acid phosphorus. A notable exception is the English potato which contains about 40–60 mg/100 g of phosphorus and about 20% of that is in the form of phytic acid phosphorus.

Tannins These are of nutritional importance in some of the millets, particularly sorghum. They are naturally occurring compounds which were first recognised in leather tanning liquors, hence the name.

Table 7.1 *Composition of some cereal foods (per 100 g)*

Food	Water (g)	Energy (kJ)	Protein (g)	Fat (g)	B_1 (mg)	B_2 (mg)	Niacin (mg)
Maize porridge (ogi)	50–80	NA	2–5	1–4	0.1[a]	0.1[a]	0.9[a]
Kenkey	60–65	560	4	0.5	NA	NA	NA
Bread (white)	35–40	990	8	1.5	0.2	0.03	1.4
Bread (white fried)	4	2300	8	40	NA	NA	NA
Rice (boiled)	70	510	3	0.1	0.01	0.01	0.8
Chapati (made with fat)	30	1380	8	13	0.3	0.04	3.4
Chapati (made without fat)	50	860	7	1	0.3	0.04	3.0
Papadams (raw)	12	1150	20	2	0.2	0.1	1.2
Cornflakes	3	1500	9	2	0.03	0.6	0.6
T'ef enjara	60	680	5	1	0.2	0.05	0.8
Arab (pitta) bread	40	1130	7	1.5	0.1	0.02	1.4

Note: NA = value not available; a = variable. Values given at 50 g water.

Table 7.2 *Total phosphorus and the percentage of phytic acid phosphorus*

Source	P (mg/100 g)	Phytic acid P (%) in total P
Wheat flour – extraction		
100% (wholemeal)	350	70
85%	270	55
80%	230	50
72%	130	30
Polished rice	100	60
Legumes	15–300	50–80
Nuts	200–500	50–80
Most fruit and vegetables	0–50	0

They bind some proteins and are widely distributed in the plant kingdom particularly amongst woody plants where it is thought they provide protection against fungi. Tannins are common in the outer layers of some sorghum varieties. This increases resistance to attack by birds but decreases nutritive value.

Toxicity of tannins has been shown in animals and man. The PER is decreased and birds fed on high tannin legumes or sorghum do not grow well. Many enzymes are inhibited by tannins. In the food industry tannins are used in wine, cider and beer production, to contribute to the flavour or aid in clarification. In sweet potato, tannin prevents oxidation of carotene, and tannins have also been added to the cooling medium to improve the flavour of smoked fish.

Tannins in sorghum have been decreased by various means before consumption. Fermentation for 48 h in the production of the Sudanese bread kissra has reduced them by 20% but the best methods seem to be parboiling after dehulling and alkali dehulling. Both reduce the level by 80%. For the latter method the grain is steeped for 5–10 min at 70°C in 20% aqueous sodium hydroxide. The grain is then neutralised with acetic acid, washed and dried.

Rice

As one might expect, the earliest records of rice are found in China, about 2800 B.C. It is next recorded in India about 1000 B.C., and reached the Middle East in 500 B.C. The first settlers to the United States took it with them in 1685, and from there it reached South America.

Rice is grown in tropical and temperate climates with abundant rain and sunshine. It requires a mean temperature of about 22°C. It is typically a swamp cereal and grown in water which serves both for irrigation and weed control. The crop is often dried before the harvest and cut by hand. In some parts of the world combine harvesters are used.

Ten per cent of the world production of rice is upland rice, which is

Fig. 7.3a. A simple rice-threshing machine in Nigeria.

Table 7.2a *Dimensions and thousand-grain weight of rice*

Grain type	Length (mm)	Length: width ratio	Weight of grains (g)
Long	6.61–7.5	3	15–20
Medium	5.51–6.6	2.1–3	17–24
Short	<5.5	<2.1	20–24

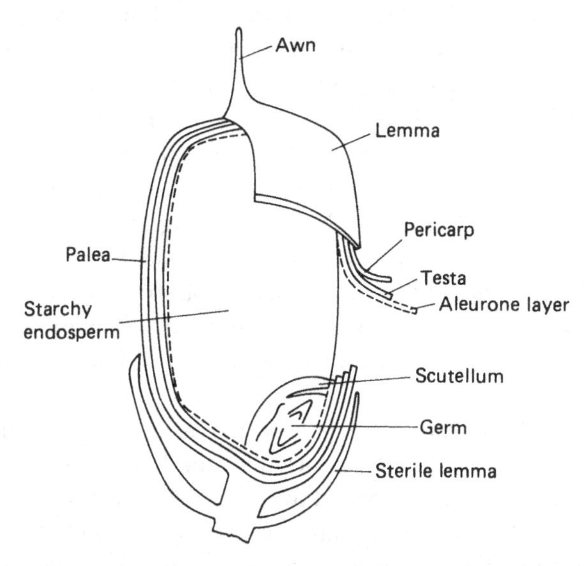

Fig. 7.4. Diagrammatic section of a rice grain.

grown in some areas of Africa and Asia on hilly ground without immersion. In some areas of the world there are two crops per annum, the main crop in the wet season, with a subsidiary crop in the dry season with irrigation. Although rice is of world-wide distribution, most of it is produced in Asia. It is the staple food of half the world's population and more than half of the crop is eaten on the farm where it is grown (Fig. 7.3a). An ear of rice is shown in Fig. 7.2, the grain in Fig. 7.3 and a section in Fig. 7.4.

In the world market rice varieties are often classified by their dimensions as shown in table 7.2a.

Most of the rice grown is consumed as food either boiled or fried, in West Africa with stew (containing meat, fish, oil and vegetables) and with fruit. In China it is eaten with pork, chicken or vegetable, in India as curry, in Spain as paella and in Italy as risotto. Some of it is parboiled, canned or used for baby foods as flour. In Japan some rice is fermented into the drink sake, to which Marco Polo already referred in the thirteenth century after his travels in China. Rice milling is considered in Chapter 13.

Maize

The American 'corn' is the only cereal of American origin. It was cultivated extensively by the American Indians. Columbus found it in Haiti where the local inhabitants referred to it as 'mahiz'. Maize requires a moderately high temperature, high moisture and loamy soil. The maize grains are born in a 'cob' which ripens in 3–5 months and therefore a long summer is required. Grain yield is higher than for any other cereal, being approximately 15 cwt/acre for most other cereals, but 22 cwt/acre for maize. Figure 7.3 shows the grain and Fig. 7.5 a diagram of it.

The grain is the largest of any cereal seed. It is flat and angular. The germ is relatively large but otherwise the anatomical structure is typical of a cereal grain. Depending again on variety, some maize grains contain predominantly a white floury endosperm, others a hard, horny or 'corneous' one.

Maize is found in three types. One, field corn, is used mainly for animal feed. Then there is sweet corn which is used for food. The surface of the grain is usually wrinkled when dry and the taste is sweet. Finally, there are 'flour' types which have softer grains and are readily milled. It is an extremely variable species and its potential for breeding new varieties has not yet been fully exploited. So far, the following varieties have been produced: hybrid maize of particularly high yield in the United States, high lysine maize which has better nutritive properties, high oil varieties yielding up to 17% maize oil, high (25%) and low (5%) protein varieties and finally high amylose (amylomaize) and high amylopectin (waxy maize) varieties.

Although maize was only introduced to Africa from Central America during the last 500 years it is now the most widely grown cereal on that continent. The cobs can be boiled whole and the grain consumed but more often the grain is milled into meal. It is then eaten as a porridge which in West Africa is usually subjected to sour fermentation. In some parts of Central America and West Africa maize dumplings are popular while in Ghana, these wrapped in leaves are the staple food, kenkey. In Central America tortillas are made by heating dried maize almost to boiling in 5% lime water for 30–50 min. The grain is then cooled, washed

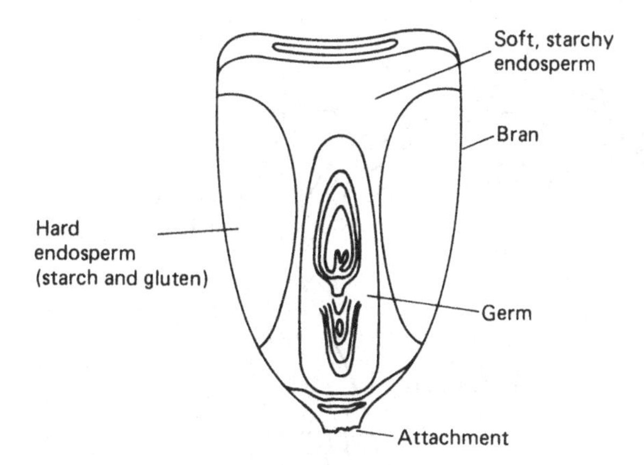

Fig. 7.5. Diagram of a maize kernel.

and drained. It is ground into a dough called masa which is shaped into pancakes and cooked on a griddle.

The treatment causes a 40% loss of carotene with yellow maize (which contains carotene, hence the colour) but a gain of 15% phosphorus and 2000% calcium. It also causes an increase in available niacin and therefore there is little pellagra in Central America. This is in contrast to maize-eating areas which do not use this lye treatment, e.g. north-West China.

Particularly in the UK and USA much of the maize is eaten as cornflakes. Some of it is canned, pickled as small immature cobs and large quantities are processed for cornflour and glucose.

The millets

These are divided into tropical and temperate types, the tropical ones being by far the most important. The grain is usually small except for that of sorghum. The tropical millets tolerate poor soil quite well and grow where little else will grow.

The most important of the millets is sorghum, sometimes called Guinea corn, kaffir corn, Indian millet, or Egyptian corn (Fig. 7.6). It probably originated in Africa where Egyptian tomb drawings show it in 2200 B.C. and Assyrian ruins in 700 B.C. Today sorghum has a wide distribution growing in Africa, India, North China, Manchuria and the United States. There are basically four groups – the grassy sorghums, sorgo (sweet sorghum with high sugar in the stem which is used mainly for forage), broom corns used for brushes and grain sorghum which is used for food.

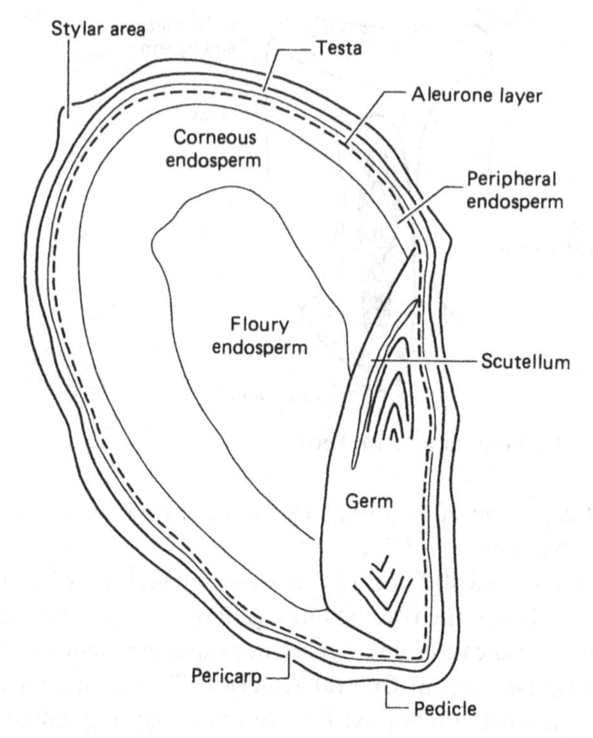

Fig. 7.6. The structure of a typical sorghum kernel.

The aspect is similar to that of maize, the plant growing 0.75–2.50 m high. Dwarf varieties are now frequently used, to allow combine harvesting.

The seed is similar to maize but rather smaller and spherical (Figs. 7.2, 7.3). There are three layers to the endosperm, each having different milling properties. The leaves are almost 1 m long and the plant carries up to 2000 seeds.

In developed countries the millets are used mainly for feed while in many areas of the world they are the food of the poor peasant. In Africa fermented and unfermented dumplings, and porridges are made as well as beer. In India millet chapatis are baked and in Asia boiled or steamed millet preparations are eaten. In West Africa sorghum is widely used for fufu (tuwo, n'dawa), fermented and unfermented porridges and fermented beers (pito). While in the United States and the United Kingdom bitterness in the beer is due to added hops, in sorghum beers bitterness is due to tannins.

Bullrush millet is one of the most drought-resistant of all millets growing in Nigeria, the Sudan, India and Pakistan. It is probably of African origin. Finger millet is probably of Indian origin. It stores well,

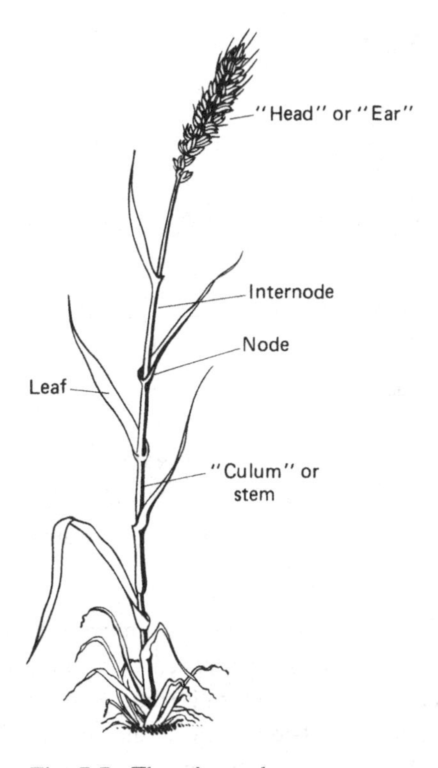

Fig. 7.7. The wheat plant.

for up to five years, and is first grown when bush is burnt in reclamation of land. It responds well to manuring and ash topping.

T'ef is a very local but nevertheless extremely important grain of Ethiopia. Its name is derived from the expression 'tefa' which means 'lost': the grain is so small that if dropped, it is impossible to find it again. (The thousand-grain weight is 0.3 g.) It is mainly made into enjara, a very thin fermented pancake but is also used for beer and spirit production.

Wheat

The bread wheat is the most important cereal of the temperate zone. The plant is shown in Fig. 7.7. The grain is elongated and has a longitudinal furrow along its length. It is rather more round and slightly larger, but otherwise similar to rice. Figure 7.8 shows a transverse and a longitudinal section of the wheat grain. The longitudinal section (A) shows clearly the outer bran coat which consists of several layers of fibre. The aleurone layer lying underneath consists of large, square cells rich in protein, vitamins and minerals. Nutritionally, this is a very important layer. The

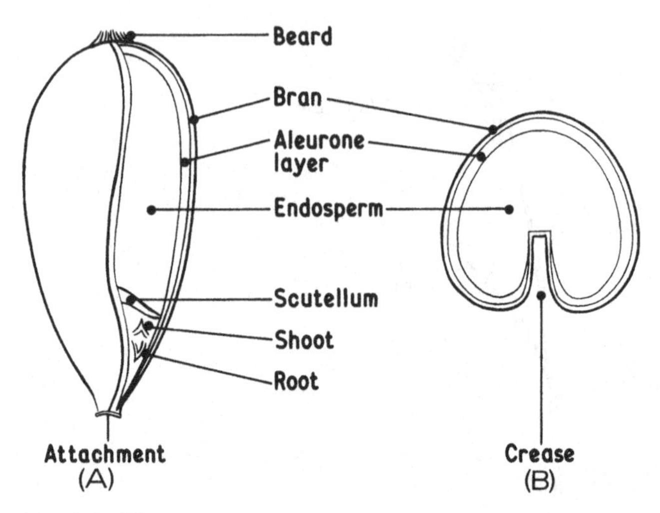

Fig. 7.8. The wheat grain in (A) longitudinal, and (B), transverse section.

endosperm below forms the food store of the germ or embryo. It is this fraction from which flour and semolina are prepared. The main component of the endosperm is starch, although there is some protein which decreases in concentration from the outside towards the inside of the grain. The most important of the protein components is gluten which is water-insoluble and of great importance in producing the dough structure which is later reflected in the structure of the bread. The scutellum (Latin, 'little shield', so-called because of its shape) separates the embryo or germ from its food store, the endosperm. The germ is fairly high in fat, vitamin and mineral content. In the transverse section (B) the typical furrowed shape of the grain is seen. The long central furrow called the crease makes the grain difficult to deal with in milling. It contains some dirt (crease-dirt) and is also the cause of separation problems. Breeding wheat grain without a crease has so far not been successful.

When ground wheat endosperm, i.e. flour, is mixed with water, the gluten forms a three-dimensional water-insoluble protein network. Both mechanical work and water are required to form a flour dough. If such a dough, after resting for 30 min in water, is washed under slowly running water, both bran and starch are washed away and the sticky gluten is left behind.

Flour strength. The term flour strength is unique to wheat. Flours may be classified into strong, medium and soft. A strong flour is one that produces large, well piled loaves. It has a high protein content of 10–12%

Fig. 7.9. Frying chin-chin (achomo). This is a popular Nigerian snack often referred to as 'African pastry'. It is prepared from wheat flour, sugar, egg and milk flavoured with nutmeg.

and a strong gluten which is very resistant to stretching. This gluten gives good support to the expanding dough on baking. It is obtained from a strong Canadian spring or similar strong wheat. Since this is fairly expensive, soft flour is often added for bread manufacture. A mixture of wheat of varying strengths is referred to as a mixed grist.

Flour of medium strength contains 9–11% of protein; the gluten is more extensible and soft. American winter, Australian, Argentinian and strong European wheat varieties fall under this heading. All-purpose household flour is made from a medium flour and 30–40% of the bread grist is made from it.

Soft flour containing 7–9% of protein is ideally suited for the manufacture of cakes and biscuits but makes poor bread. The gluten is soft and extensible and there is little of it. Soft Australian and European flours are of this type. It is worth noting that US and Canadian wheats are well standardised for export purposes. European wheats are usually of very variable quality due to the small fields and rather poor grading systems. In developed countries wheat is usually harvested with 'combine harvesters'. These are large machines which are driven across the fields and which cut the wheat, thresh it and pour the grain straight into a truck which runs alongside.

Triticale

This is the first man-made cereal, a hybrid of wheat (*Triticum*) and rye (*Secale*). It was first grown commercially in the United States in 1970 and US grain standards were produced in 1977. Triticale grows well in light, sandy and acid soils. In breadmaking quality it is not equal to wheat but better than rye. It produces a large variety of good quality cereal products such as pasta, biscuits and chapatis.

Legumes

The legumes are a very successful group of flowering plants. They are of world-wide distribution and every area has its local 'pea' or 'bean'. Groundnuts, French, Lima and runner beans are found in South America, soya beans in the United States and in the Far East, the cowpea and broad bean in Africa and lentils and chick peas in India. The fruit is a pod containing 2–10 seeds (Fig. 7.10).

As regards composition, there are basically two groups of legumes. First there is the high-protein high-oil group. This comprises soya, groundnut, lupin and winged bean. These legumes are mainly used for

Fig. 7.10. The garden pea. The lower picture shows an opened pod containing seven seeds.

processing. Protein content is as high as 35% and oil content varies from 15 to 45%. The second group comprises the moderate-protein low-oil types. The representatives of this group are most important as human food. Examples are the cowpea, gram, pea and lentil as well as the phaseolus group. The latter includes the lima bean, cow pea and bambarra groundnut (Fig. 7.11). Protein content is of the order of 15–30% and oil content 2% or less. In contrast to the high oil varieties, starch is often present. Moisture content lies between 5 and 10% and the seeds store well. An additional bonus with legumes is the fact that they harbour symbiotic bacteria which fix nitrogen from the air.

As a group, the legumes are high in B vitamins but low in the sulphur-containing amino acids methionine and cystine. They are high in lysine. This is unusual for plant proteins and therefore legumes form the ideal additive to cereal diets, the cereals being low in lysine (see composite flours, p. 119). As in the case of cereals, the legumes tend to have a high yield at the cost of a low protein content and vice versa. It has been suggested that a low content of methionine and cystine is also related to high protein content. It would appear that the storage proteins of the legumes are low in the sulphur-containing amino acids but not the structural proteins.

Fig. 7.11. Some common fruits and seeds. Top row, left to right: oil palm fruit (section), fruit (side view), blackeyed peas, calabar beans. Middle row: cashew nuts, groundnuts, soya beans, cottonseed. Bottom row: cocoa beans, melon pumpkin seeds, oilnut.

As a group, the legumes have certain nutritional disadvantages. Apart from their methionine and cystine deficiency, their digestibility appears to be low and a high proportion of legumes in the diet causes flatulence. Another significant disadvantage is their toxicity. Legumes may contain toxins or chemicals which inhibit enzyme reactions.

If butter beans are steeped in water in a closed vessel overnight the bitter almond smell of hydrogen cyanide is easily perceived. Some legumes contain haemagglutinins which cause damage to red blood cells and intestinal cells. Others cause haemolytic anaemia (favism) ('haema' is Greek for blood and 'lysis' for to release or dissolve) or interfere with collagen formation. An important disease is lathyrism. This is caused by a species of tare (*Lathyrus sativus*). If eaten in small amounts the seeds are a useful food but if they become a staple the disease is caused. In parts of India the seeds, which are resistant to drought, are often sown in case the

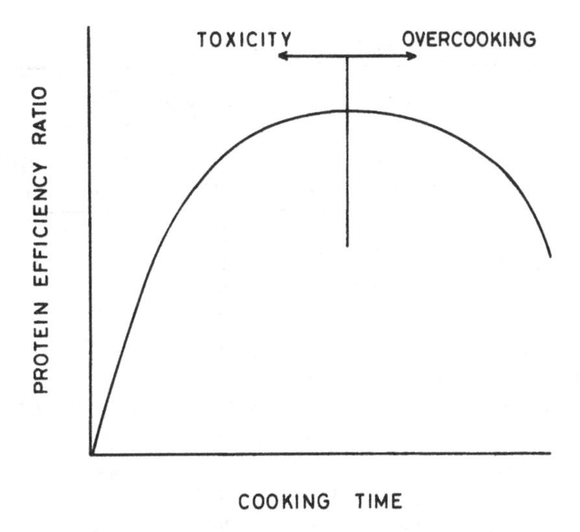

Fig. 7.12. The relation between protein efficiency ratio and cooking time for legumes.

wheat harvest fails. If it does an epidemic of lathyrism may occur and cause in its victims very severe paralysis of the legs. The toxin can be extracted by heating the seeds in four volumes of water for one hour.

Other toxins of legumes are destroyed by soaking for 24 h followed by extensive boiling. There is an optimum cooking time, because on overcooking the protein efficiency ratio tends to decrease. This is shown in Fig. 7.12. The composition of some legumes is given in Table 7.3.

Soya

The soya bean probably originated in the Far East. The seeds contain about 10% moisture, have a high protein and oil content of about 35% and 15–20% respectively, and contain 5% fibre and 5% ash but no starch. The PER is 2.32. The calcium content is relatively high (0.25%) and the thiamin content 12 μg/g. However, the latter is lost on heating.

A simple treatment of soya beans at village level consists of washing the beans and soaking them in cloth bags in water at 27°C for 7 h. The beans are then boiled at atmospheric pressure, sun-dried and cracked in a hand-operated attrition mill. The grits are then ground into flour. The heat treatment easily inactivates anti-nutritional factors. For instance, untreated meal has a PER of 1.3, which after 15 min steaming is raised to between 1.9 and 2.0. Trypsin inhibitor activity is also reduced by more than 95%. However the heat treatment tends to render protein insoluble and darken the colour of the meal.

Table 7.3 *The composition of some legumes (per 100 g)*

	Water (g)	Energy (kJ)	Protein (g)	Fat (g)	Total carbo-hydrate (g)
Bambara groundnut	10	1550	20	6	60
Blackeye pea (raw) (1)	13	1360	23	1.5	58
(decorticated) (2)	30	1080	20	1.2	44
(moi-moi) (3)	70	500	7	2	20
(akara) (4)	60	920	9	15	14
Black gram (urad dahl)	10	1450	24	1.5	60
Chick pea (Bengal gram)	11	1360	20	5.6	60
Humous (chick pea paste)	60	770	8	13	11
Groundnut (dry)	5	2570	26	48	20
Lentil (masur dahl)	12	1300	20	0.6	65
Lima (Butter) bean	12	1162	20	1.5	58
Mung bean	12	980	22	1.0	35
(sprouts)	90	146	3.8	0.2	6
Pea	78	280	5.8	0.4	11
Soya bean	10	1700	40	20	20
Winged bean	10	1700	33	17	37

Notes: (1) Cowpea also called 'bean' in Ghana. (2) Beans soaked in water, shells removed. (3) Decorticated beans mashed with water, onion, salt and oil. Steamed. (4) Decorticated beans mashed with water, onion, salt. Deep fried.

Groundnut (peanut, monkey nut)

This is closely related to the pea or bean (Fig. 7.13). It comes originally from Brazil or Peru in South America. It was taken to Africa by the early explorers and the traders brought the groundnut back to North America.

One of the biggest producers in the world is Nigeria and very large quantities are exported (Fig. 7.14).

After fertilisation the flower stalk elongates, forcing the pods underground where they develop. In the erect varieties the pods remain close to the stem and are less scattered than in the prostrate varieties. Hence, the former are easier to harvest. Harvesting is normally by machines in the United States, but in other parts of the world the crop is dug up with a spade.

As regards composition fresh seeds have a moisture content of 45% which is rapidly reduced to 5%. As with the soya bean, oil and protein content are high, about 50% and 25% respectively (see Table 7.4). PER is 1.65. There are 2% crude fibre, 2% ash and a little starch, usually less

Fig. 7.13. Groundnuts: top showing nuts in their outer shell; middle and bottom are surface views.

than 5%. Groundnuts are also good sources of riboflavin, thiamin and nicotinic acid, although the thiamin is usually destroyed during heat treatment.

In West Africa an important use of groundnuts is in cooking. Here maize and groundnuts are either roasted together and eaten as such, or the groundnuts may be used in soups and stews. They are broken up first and boiled in the soup, together with meat, fish and vegetables. Groundnut stews are also popular in East Africa where they are consumed after

Table 7.4 *Composition of groundnuts and some of their products (per 100 g)*

	Water (g)	Energy (kJ)	Protein (g)	Fat (g)	Total carbo hydrate (g)	Fibre (g)	Ash (g)
Raw seeds (dry)	5	2570	26	48	18	2	2
boiled	37	1570	16	32	15	2	2
roasted	2	2630	26	49	21	2	2
Peanut butter	1	2425	25	50	15	2	4
Defatted peanut flour	8	1550	50	10	30	3	4

Fig. 7.14. Groundnuts are one of the main exports of Nigeria. (Courtesy Unilever plc.)

first roasting and then pounding into a paste. This may be either a coarse or a fine paste, similar to the peanut butter eaten in the West. After soya and sunflower seeds, groundnuts are the third largest source of vegetable oil in the world.

In the manufacture of peanut butter, the beans are roasted, cooled, blanched and ground. The material is then salted using salt and sugar, finely ground in order to avoid grittiness.

Salted groundnuts are first submitted to steam blanching and brushed to remove the skin. There are several types of blanching – wet, dry, hydrogen peroxide and alkaline bleaching. The object is to bleach and remove the skin. The salt is usually added by dispersing it in alcohol-soluble zein (maize protein) to make it adhere to the groundnut.

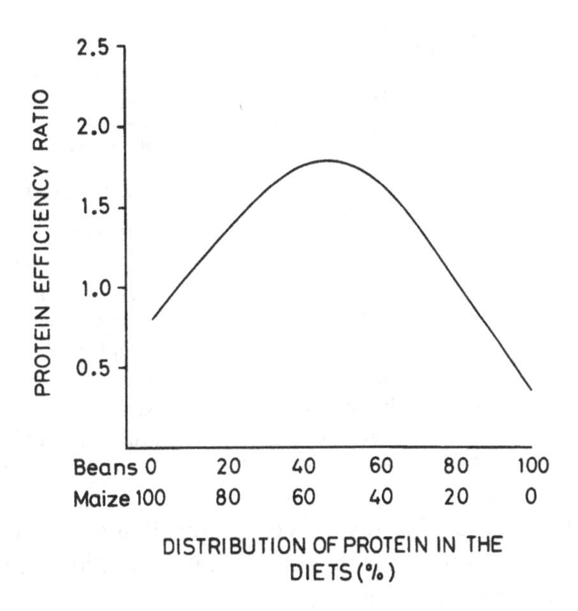

Fig. 7.15. The effect of the bean : maize balance on the protein efficiency ratio.

Composite flours

It has been pointed out (p. 113) that many legumes tend to be deficient in the sulphur-containing aminoacids, but cereals contain adequate levels of these. The cereals in their turn are low in lysine. Therefore cereals and legume mixtures supplement each other's deficiency in 'mutual supplementation' (Fig. 7.15). The principle of combining cereals with legumes has been used for centuries. In India dahl (legume) porridge is eaten with wheat chapatis. In the Middle East sesame, chick pea and wheat have been used to make Arab bread and lentils together with cereals to make gruel. In the southern US maize grits and cow peas are eaten together, and in Latin America frigoles (beans) are used with maize tortillas.

In the 1950s the Institute for Central America and Panama (INCAP) developed a maize–cottonseed meal referred to as 'incaparina'. This was followed by corn–soy milk (CSM) and wheat–soy blend (WSB) developed in the US. Many other composite foods have been produced and some are consumed in developing countries in very large quantities (Table 7.5). Such products are cheaper than most animal proteins and can be specially tailored for a given population group such as children, nursing mothers or the aged. No cold storage is required and the foods are

Table 7.5 *Some examples of composite flours*

Country	Product	Components
Brazil	Ceralina	Maize, soya
Columbia, Guatamala Panama	Incaparina	Maize, cottonseed
Ethiopia	Paffa	Teff, soya
India	Bal Ahar	Maize, cottonseed, groundnut
Lebanon	Laubina	Wheat, chick pea, milk powder
Nigeria	Soy-ogi	Maize, soya
Philippines	Nutripak	Rice, fishmeal
Senegal	Ladylae	Millet, groundnut

quickly prepared for use. However, composite flours are often unfamiliar, and more expensive than the cereal staple. Nutritionally they are not as good as a proper mixed diet.

8

Fruits and vegetables

In botany a fruit is a structure derived from the fertilised ovary of a plant and which serves to reproduce that plant. In food science a fruit is the edible part of a plant normally eaten raw. This distinguishes a fruit from a vegetable which is normally eaten cooked even if botanically it is a fruit such as okra, peppers or pumpkin.

On ripening the fruit loses chlorophyll and develops other colouring pigments. The size of the fruit increases and it changes in flavour, texture and nutrient composition. After the fruit is fully ripe it begins to deteriorate, becomes less attractive and the content of vitamin C decreases.

Although some fruits such as peaches, apricots, mango and papaya may contain high levels of carotene, the main nutritional use of fruit is normally for their content of vitamin C. Protein content is usually below 1%, fat and ash below 0.5% and carbohydrate generally less than 10%, up to 3% being crude fibre. Water content is usually high, of the order of 80–90% and energy content accordingly low (80–200 kJ/100 g). Important exceptions are banana, plantain, matoke and dates which are used as staples (Table 8.1).

Green vegetables, although eaten the world over, are never a staple of the diet because leaves, like most fruit, do not normally contain storage nutrients such as starch, protein or fat. They contain high levels of water (80–90%) and are characterised by protoplasmic rather than storage protein. The protein content is of the order of 1–4% and rarely as high as 5 or 6%. Fat content is 0.1–0.3%, total carbohydrate 3–10%, of which 0.5–1.0% is crude fibre. Ash content is usually less than 1%. Nutrient composition is affected extensively by light, season, location, fertilisation and genetic characteristics of the plant.

Leafy vegetables, beause of their large surface area, tend to wilt rapidly particularly at high temperatures and low relative humidities. This

Table 8.1 *Composition of some raw fruits (per 100 g)*

Fruit	Water (g)	Energy (kJ)	Protein (g)	Fat (g)	Total carbo-hydrate (g)	β-carotene equiv. (μg)	Vitamin C (mg)
Avocado	75	700	2	15	6	300	18
Banana	75	355	1	0.2	22	100	10
Breadfruit	70	430	2	0.3	25	NA	NA
Guava	83	259	1	0.5	15	300	300
Lime	85	208	1	0.2	12	0	40
Mango	80	276	1	0.5	17	3200	40
Orange	86	150	1	0	8.5	230	50
Papaya	75	353	5	1	17	1000	50
Pineapple	85	218	0.5	0.2	14	100	30
Tomato	93	60	1	0	3	600	20

NA = Not available

presents a problem in transportation. Green vegetables are an important source of vitamin C and this may suffer extensive losses during wilting. In one experiment, 40% of vitamin C was lost in kale during storage of three weeks at 0°C, four days at 10°C and two days at 21°C.

While riboflavin is relatively stable, significant losses of β-carotene may occur. Leaves commonly eaten in the tropics are: bitter leaf, sweet potato tops, and the leaves of garden egg, pumpkin, okra, cocoyam and very many others. If the leaves are very dark green (e.g. spinach, sweet potato tops) they tend to contain high levels of β-carotene. The paler cassava and pumpkin leaves contain much less and very pale leaves may contain hardly any (See Table 2.9, p. 39). Reduction in storage temperature will reduce both wilting and nutritive loss. Table 8.2 gives the composition of some vegetables.

Fruits as staples

Banana, plantain and matoke

The banana belongs to the botanical genus *Musa*. It is a giant herb up to 10 m high. What appears to be the stem are really overlapping leaf bases. The leaves are very large, succulent and tattered. Vegetative propagation of the plant is by suckers. Within one year the plant flowers, having the male flowers at the bottom and the female ones at the top of the bunch.

Table 8.2 *Composition of some raw vegetables (per 100 g)*

	Water (g)	Energy (kJ)	Protein (g)	Fat (g)	Total carbo-hydrate (g)	β-carotene equiv. (μg)	Vitamin C (mg)	Iron (mg)
Peppers	90	65	1	0.5	5	10	30	0.1
Okra	90	70	2	0.3	6	150	50	1.0
Pumpkin	90	70	1	0	8	350	15	1.0
Garden egg	95	60	1	0	5	0	5	1.0
Onion	90	100	0.5	0	1	0	3	0.1
Leaves	80–95	20–200	1.5	0.2–1.0	4.5	10–200	10–300	0.1

Fig. 8.1. The fruit of the papaya or pawpaw tree is green or yellow, with yellow-orange flesh. It is the source of the proteolytic enzyme papain used as a meat tenderiser.

Fig. 8.2. Durian is a large fruit weighing 3–4 kg. It is dull yellow and covered in spines, popular in South-east Asia.

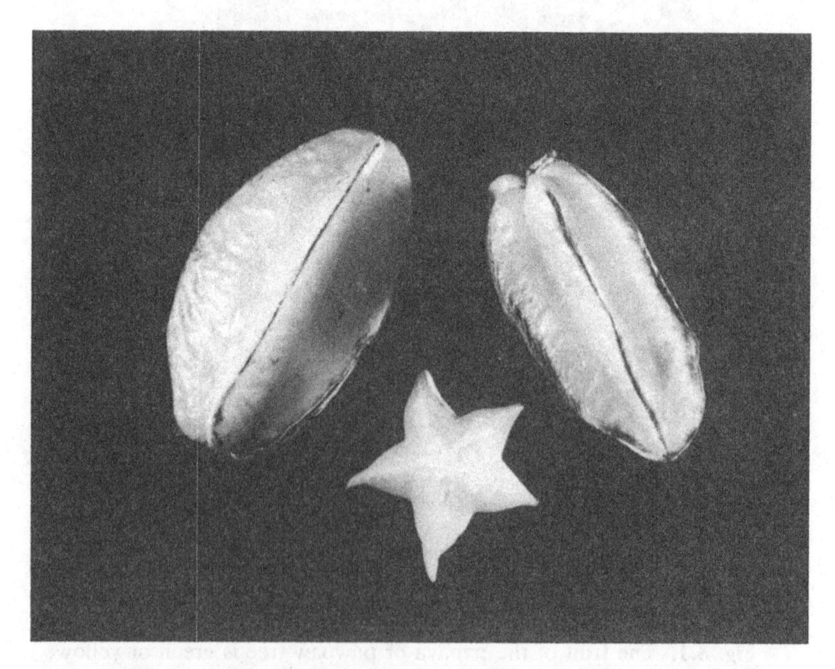

Fig. 8.3. The star fruit, 7–12 cm long, comes from Indonesia. It is used for fruit salad, jam and in making a drink.

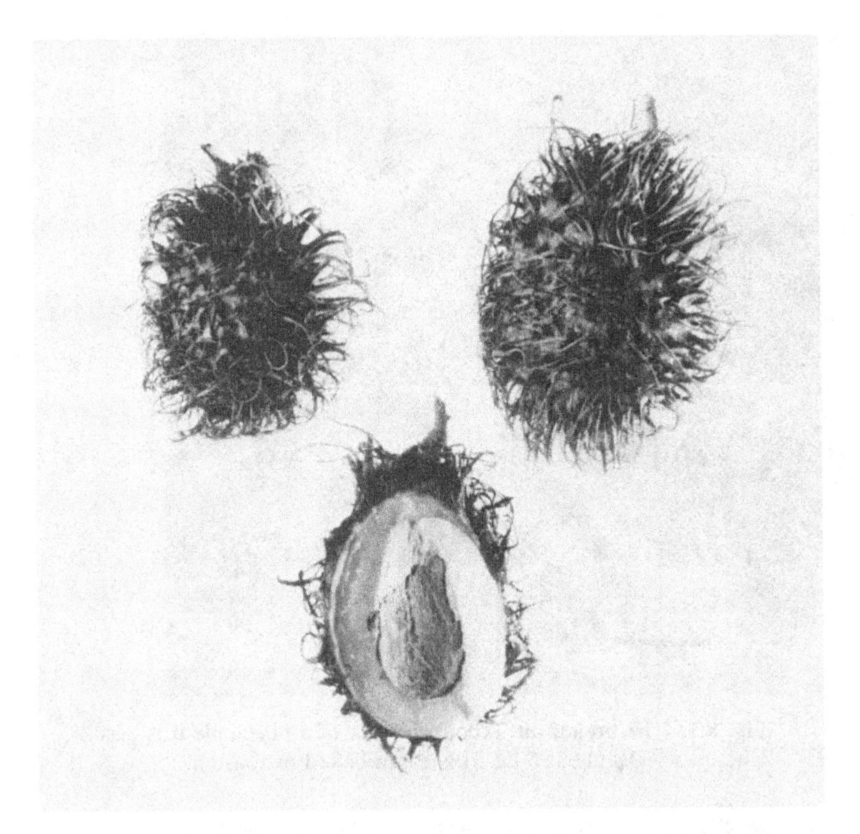

Fig. 8.4. Rambutan is a fruit of South-east Asia and similar to litchi. The fruit is the size of a plum. It is the white flesh surrounding the single seed which is eaten.

The fruit is arranged in hands of 12 to 16 fingers. There are approximately 9–12 hands per bunch, giving rise to between 100 and 200 bananas per bunch (Figs. 8.7, 8.7a).

After fruiting the stem is cut down and new suckers are thrown up. During ripening the fruit changes colour from green to yellow. This colour change is accompanied by a sweetening of the fruit caused by hydrolysis of the starch and increase in sugar concentration. This is shown in Table 8.3.

The export trade of bananas is well organised, the largest exporter being Ecuador to the United States. Europe obtains most of its bananas from West Africa. For transport, either bunches or fingers are despatched naked or covered with paper, straw or polythene. Ripe or damaged fruit is rejected. In the producer country bananas are left to ripen by themselves, but for export, ripening occurs under strictly

Fig. 8.5. The breadfruit. About the size of a pineapple it is best when it weighs about 5 kg. It is eaten baked or roasted.

Fig. 8.6. The jak or jackfruit is related to the breadfruit. The fruit is very large and can weigh up to 70 lb. It is popular in Asia and may be eaten cooked or raw. (Courtesy R. E. Muller.)

Fig. 8.7. When bananas are a staple, they are often eaten together with cassava. As a result protein intake may be too low.

controlled conditions. On the ship the bananas are first stored at 11–13°C and on unloading the temperature in the store is raised to 15°C. The fruit is kept at that temperature until the ripening procedure is initiated. This is set off approximately four days before sale to reach colour index 3–5 (see Table 8.3). The temperature is raised to 20°C without ventilation. If desired, ethylene gas which accelerates ripening, is now injected into the chamber. The fruit is then left for 15 h. The temperature is now reduced in stages to 16°C with occasional changes of air until the fruit is ripe (colour index 3–5).

Table 8.3 *Ripening changes in bananas*

Colour index...	1	2	3	4	5	6	7
Peel colour...	Green	Green trace yellow	More green than yellow	More yellow than green	Yellow-green tip	All yellow	Yellow flecked with brown
Average sugar concentration (%)	1	4	5	10	14	18	18
Average starch concentration (%)	20	18	16	12	6	3	2

Fig. 8.7a. Banana market in Djakarta (Indonesia). (Courtesy F. A. Leeming.)

The fruit of plantain is very similar to that of the banana except that it is less sweet and rather larger, being between 30 and 45 cm long. It has a higher starch and lower sugar contact. Plantain is often referred to as cooking bananas and is a chief staple in many parts of Africa, particularly East Africa. It is one of the minor crops in West Africa and is usually eaten roasted, fried or boiled. The analysis of banana, plantain and some of their products is given in Table 8.4.

Table 8.4 *The composition of banana, plaintain and some of their products (per 100 g)*

Food	Water (g)	Energy (kJ)	Protein (g)	Fat (g)	Total carbo-hydrate (g)	Ash (g)
Banana						
ripe	70	340	1	0.3	20	1
powder	3	1120	4	1	64	3
Plantain						
ripe	70	500	1.3	0.1	30	1
boiled, ripe	70	480	1.1	0.1	28	0.5
fried, ripe	47	930	2	5	45	1
roasted, ripe	56	695	2	0.2	41	1
roasted, unripe	42	910	2	0.2	54	1

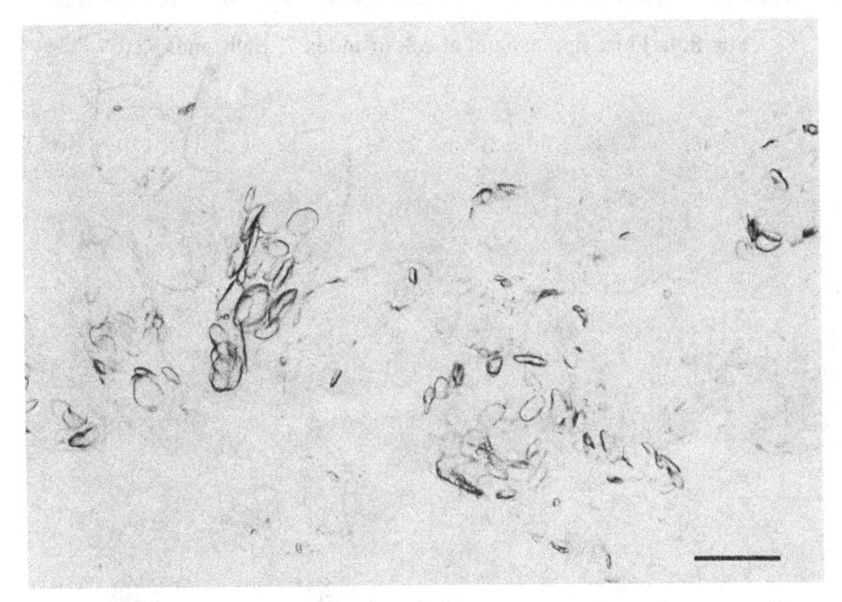

Fig. 8.8. Cells of banana pulp. Some of them are filled with starch granules. Scale bar = 100 μm.

Matoke is similar to plantain but much smaller and remaining green. It probably originated in the West Indies or Central America where it is found today. It is a staple in parts of East Africa.

Fig. 8.9. Left: ripe banana at colour index 7; right: matoke.

Fig. 8.9a. The pineapple is perhaps one of the best known fruits of the tropics. When ripe the flesh is soft and sweet but when canned it is crisper and less sweet. The fruit is a multiple organ formed by the coalescence of smaller fruitlets around a central core. There are useful amounts of vitamins A and C in it.

Fig. 8.9b. This market stall on the University campus at Ibadan shows a fine display of oranges, coconuts and pineapple.

Fig. 8.9c. A fruit seller in a market in Malaysia. Mango (front of table) comes in many shapes: pear-shaped, elongated, oval and round. It is perhaps one of the most prized of all tropical fruits. (Courtesy H.J.S. Taylor.)

Fig. 8.10. The date palm showing bunches of fruit.

Fig. 8.10a. A good bunch of dates may consist of 40 strands each with 25–30 dates per strand. (Courtesy T. R. Hall.)

The date palm

Mainly a subtropical palm which attains perfection in the warm and dry zone which stretches from Senegal to the Indus i.e. between the 15 and 13 degrees of latitude. In humid climates the fruit is of poor quality. The tree bears fruit about four to five years after planting. The dates are borne in bunches consisting of about 40 strands each with 25–30 dates. Average yield is 50–75 kg per tree per year. Ripe dates contain 20% water, 3% protein and 0.2–0.3% fat. Carbohydrates are mainly glucose and fructose with up to 2% sucrose. Some varieties contain up to 20% sucrose (Figs. 8.10, 8.10a).

Oilseeds and fruits

Oil is obtained commercially from quite a number of plant sources. Sometimes the oil is a byproduct (e.g. as with maize or cottonseed) and at other times the oil is the main product (e.g. sunflower, oil palm). Table 8.5 lists the most important sources of plant oils with their percentage oil content (at natural moisture content).

As in animals, oleic (18 : 1) and palmitic (16 : 0) acids are prominent in plants. However, the latter also contain linoleic acid (18 : 2), an essential fatty acid, which is one of the most common constituents of seed oils. The exception is coconut oil which contains mostly saturated fatty acids but little oleic and essential fatty acids. Thus basically a diet of vegetable seed oil is rich in linoleic and to a lesser extent linolenic acid. Melon seed (Egwusi) contains erucic acid which is toxic.

Oil palm

This is one of the world's most important source of oil for food and soap manufacture. The plant gives a higher oil yield per acre than any other. It originates in West Africa where today it still grows wild.

Table 8.5 *The most important sources of plant oils (natural moisture content)*

	Oil (%)		Oil (%)
Maize	5–10	Olive	15
Groundnut	40–45	Sesame	45–50
Soya bean	15	Sunflower	30–40
Cottonseed	20–25	Oil palm	45
Melon seed	45	Coconut flesh	65
Rapeseed	35–40	Castor bean	45–50

Fig. 8.11. The oil palm showing bunch of fruit.

Oil-palm plantations are common in the Congo, Ivory Coast, Indonesia and Malaysia. The plant favours a tropical climate with relatively high rainfall. Female and male flowers occur on the same tree and the female flower gives rise to a bunch of about 200 fruits. Every year there are about 2–6 bunches on each tree. Figure 8.11 shows the bunch on the tree and Fig. 8.12 two detached bunches. In Fig 7.11 a fruit is shown both in surface view and in section.

The fruit, about 4 cm in length, is covered by a skin with the oil-bearing fibrous pulp below. This in turn contains a hard and fibrous shell, or 'nut'

Fig. 8.12. Two oil-palm bunches. (Courtesy Unilever plc.)

Fig. 8.13. Transportation of oil-palm fruits by buffalo and specially designed cart in Malaysia. (Courtesy Unilever plc.)

which is often used as fuel. Inside this shell the palm kernel is found. The oil of the outer fibrous pulp is referred to as palm oil and that of the palm kernel as palm kernel oil. The proportion of these two oils depends on the species. The extraction of palm oil is considered in Chapter 13. This oil

Fig. 8.14. The coconut palm.

contains very high amounts of β-carotene (it is the world's richest plant source) but repeated heating will destroy the vitamin as can be seen by the loss of the red colour. In tropical temperatures palm oil is semi-liquid and on standing separates into equal volumes of a dark-red fraction of high carotene content and a yellow solid fraction.

The coconut palm

This is typical of tropical lowlands. The trees are about 25 m high and bear the coconuts in bunches. Each nut has a hard shell with a layer of white

Fig. 8.14a. The coconut palm plays a very big part in the life of many people. The fruit provides food and drink, the shell serves as a vessel, the fibre is used for rope and matting, the tree trunk for house building and furniture and the leaves for thatch. The copra is a valuable export.

meat on the inside. When unripe the nut contains coconut milk – a pale whitish liquid with a strong taste of coconut. This liquid is gradually absorbed as the nut ripens. The coconut meat is either sun-dried or kiln-dried and referred to as 'copra'. Its oil content is very high, of the order of 60–65%. To extract it the fibrous husk is first removed by hand. The nut is then split into two halves and the flesh dried in the sun (Fig. 8.14a). This takes about three days. Usually fat is extracted in two stages. First there is light pressing with a single-stage expeller. Then the remaining cake is reground and extracted with a solvent or a hydraulic press. The maximum commercial fat yield for dried copra is about 60%. Coconut milk contains 95–99% water. It can be spray-dried (Chapter 15). The name of the nut is derived from the Portuguese word 'Coco' which means 'ugly face' because the marks at the broad end of the nut seem to resemble the face of a monkey.

Cottonseed

When the cotton is grown for fibre the seed is separated from the fibre at the ginning stage and may be pressed to produce an oil. Since cottonseed

Fig. 8.15. Cottonseed is a major source of both oil and protein. (Courtesy Unilever plc.)

is produced together with another important material, the fibre makes it unusual amongst the oil seeds and fruits. The seed available depends to some extent on the market for the fibre (Figs 7.11, 8.15).

The oil is particularly rich in linoleic acid and is used in the manufacture of margarine, shortening and salad oil. The press cake is an important source of protein. However a problem with utilisation is the presence of gossypol. This is a highly reactive substance contained in small glands visible under a hand lens as dark specks. It is toxic but is removed during oil processing.

Root crops

Storage roots and tubers have a relatively high water content of about 75%. Compared with cereals or legumes this results in higher transport costs, shorter shelf life and less nutritive value per unit of raw weight. Root crops are also easily damaged during transport and storage which results in losses through rotting and sprouting. For these reasons, root crops are not usually exported in large quantities.

Table 8.6 *Composition of some root vegetables (per 100 g)*

	Water (g)	Energy (kJ)	Protein (g)	Fat (g)	Total carbo-hydrate (g)	Ash (g)
Yam						
raw	75	400	2	0.2	25	1
flour	10	1440	7	1	80	4
boiled	73	425	2.5	0.1	23	1
shallow-fried	65	630	4	5	24	1
deep fried	50	890	4	7	36	1.5
roasted	60	600	3	0.5	33	1.5
fufu	71	455	1	0.2	27	0.3
Cassava	60	640	1	0.2	37	2
Cassava flour	10	1500	8	1	87	3
Gari	15	1330	1.5	0.5	80	1
English potato	75	320	2	0.1	20	1
Sweet potato	70	420	1.5	0.3	26	1
boiled	66	530	2	0.3	30	1
deep fried	27	1240	5	6	60	1

The storage material of root crops is usually starch. Protein content is generally low at approximately 3%, although new varieties of potato have been bred containing up to 10%. Protein deficiency diseases may occur if roots and tubers rather than cereals or legumes make up the basic diet. There are a large number of root crops which are used as vegetable adjuncts such as beetroot, radish, carrot, swede, turnip and Jerusalem artichoke. Sugar beet is an important source of sugar in colder climates. The most important root crops of the tropics are yam, cassava, cocoyam and sweet potato. An analysis of these and some of their products is given in Table 8.6.

Yams

These belong to the genus *Dioscorea* which contains approximately 600 species. The plant produces one or more swollen underground stems called rhizomes. It is a climbing plant with heart-shaped leaves. The tubers are large and often weigh between 10 and 15 kg and weights up to 50 kg have been recorded. The flesh is white or yellow and covered by corky outer layers. One plant may give rise to either one large or several smaller tubers depending on variety (Figs. 8.16, 8.16a).

The biggest producer of yam is Nigeria followed by the Ivory Coast,

Fig. 8.16. The yam is a climbing plant.

Togoland and Indonesia. Yields are about 3–10 tons per acre and yam is often used as a convenient food store because it may be left in the ground until the next planting season. All species apparently originate in the Old World except for cush-cush yam (*Dioscorea bifida*) which comes from America. Most species are tropical but two species (*D. opposita* and *D. japonica*) grow in Northern China and Japan. The optimum growth temperatures for the tropical varieties lie between 20° and 30°C and they require high moisture during the wet season.

The processing of yam is similar to that of potato. The tubers are

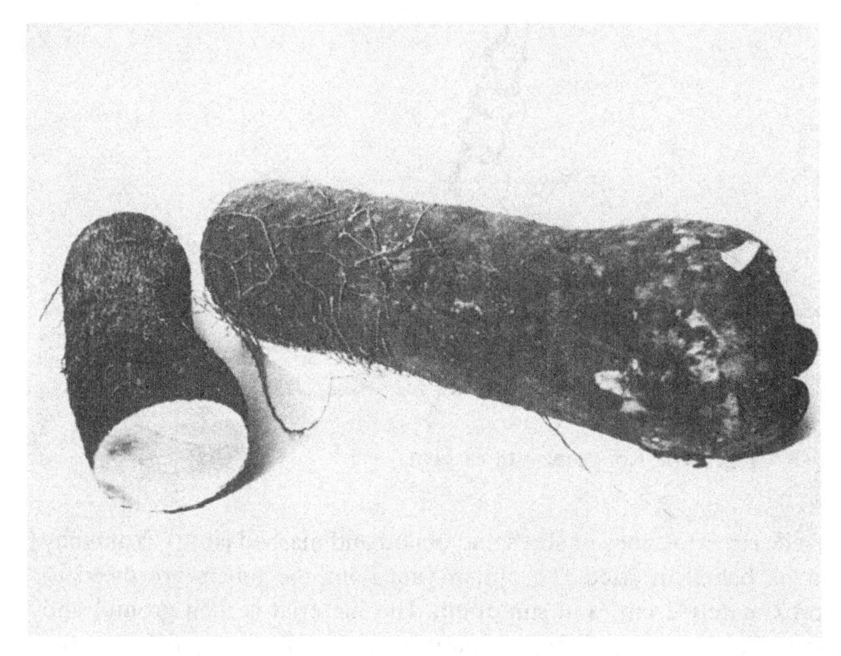

Fig. 8.16a. The West African Yam Zone is the most important yam-producing area in the world. Nigeria provides about half of the world's crop. (Courtesy E. Ossai.)

Fig. 8.17. Yam in a West African market.

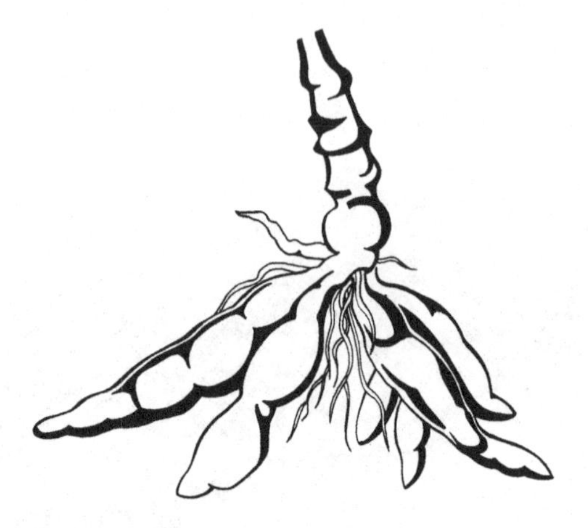

Fig. 8.18. Root tubers of cassava.

peeled, cut into cubes or slices and boiled and mashed (fufu). Yam may also be baked or fried. To obtain yam flour the tubers are diced to approximately 1 cm^3 and sun-dried. The material is then ground and stored. Before eating it is boiled in water and is easily distinguished on the plate by its brownish black colour in contrast to fresh mashed yam which is pale yellow or white. Yam can also be flaked by roller drying to a moisture content of 3–5%. A small amount of yam starch is also produced.

Cassava

Belonging to the genus *Manihot utilissima* cassava is often referred to as manioc. The plant forms large edible tubers. They are cheap to grow and its use will increase in the future. They are somewhat smaller than yam and more elongated (see Fig. 8.18). There are both sweet and bitter varieties of cassava which can be left in the ground for up to three years without attention but once lifted they deteriorate rapidly.

An important problem with cassava is its toxicity in the bitter varieties. This is due to the presence of compounds which give rise to hydrogen cyanide, extremely toxic to most forms of life. Deaths have occurred from the consumption of poorly prepared cassava and regular eating can cause chronic illness if the tubers are not prepared properly. The cyanide is bound to a glucose sugar molecule and must be eliminated. Either extensive soaking which gives rise to enzymic breakdown is used, or the grated tuber is allowed to ferment and is then heated. Steam and the cyanide escape into the atmosphere.

Fig. 8.19. Hand peeling of cassava in Nigeria. (Courtesy I. Nkama.)

Processing varies in different parts of the world and also depends on the local state of technology. In West Africa gari is a popular food. To prepare it the tubers are peeled, washed and grated. The material is then placed into sacks and fermented for from three to four days under pressure. This process removes hydrogen cyanide and develops the flavour. The cake is then broken up by sieving and heated (often with a little palm oil) to drive off water and cyanide. The product is then cooled and stored in sacks. It is consumed dry, mixed with cold water, or mixed with hot water and pounded. Modern methods of gari production have

Fig. 8.20. Peeling of cassava in Nigeria using a rotating abrasive drum. (Courtesy I. Nkama.)

Fig. 8.21. Grating cassava before fermentation in the production of gari (Ghana).

been partly mechanised. Fermentation may be reduced to two days if a starter is used (Figs. 8.19–8.27).

In South America cassava is the staple food of the Amerindian people. After peeling and grating the cassava is pressed in a matapee. This is a

Fig. 8.22. Sacks of grated cassava in the production of gari (Ghana).

basketwork tube (Fig. 8.24) unique to South America. It can be made short and fat for packing the grated cassava inside and is then stretched long and thin to squeeze out the juice.

The damp meal is then dried on wooden shelves over a fire. The meal can then be baked into round flat loaves, boiled as dumplings or fermented into an alcoholic drink.

The expressed toxic juice sprinkled on water has been used to stun fish (this is now prohibited) and as a wash to rid dogs of their fleas. When the

Fig. 8.23. In Ghana heavy stones are placed on the sacks of grated cassava to express the juice.

juice has been detoxicated by extensive boiling a savoury dip for cassava bread can be made. Further boiling produces a thick liquid called cassareep. This is a preservative used in the 'pepperpot', a dish popular throughout South America and the Carribean.

More advanced methods of casava utilisation are also available in some areas.

The tapioca known in the West is almost pure cassava starch prepared without fermentation. A wet cassava starch paste is pressed through a sieve and partially gelatinised on a hot surface. It is finally dried to a moisture content of about 12%. The preparation and analysis are virtually identical to those of sago obtained from the pith of the sago palm.

In India and the Philippines a rice substitute is prepared. The grated tubers are pelleted, steamed and sun-dried. They are then used like rice.

Extensive processing of cassava is only required with bitter varieties, i.e. those which contain significant amounts of cyanide. The lethal oral dose for acute cyanide poisoning is 0.5–3.5 mg/kg of body weight. Alkaline cyanides are approximately twice as toxic as hydrogen cyanide. With large doses of cyanide death occurs within a few minutes. Survivors

Fig. 8.24. The traditional method of pressing cassava in Central America is in a matapee, a plaited basket-work tube (see text).

have reported mental confusion, stupor, convulsions and coma. With smaller doses there is headache, a tightness in the chest and lassitude.

There is also a suspicion that chronic poisoning might be caused where cassava forms the staple diet. It is possible that a consistent and relatively large intake of protein protects against it. This may be connected with a removal of the sulphur-containing amino acids which has been observed with chronic poisoning. Proteins would supply these amino acids. Another chronic effect of cassava toxicity may be endemic goitre found particularly in Zaire. It is thought that some cyanide gives rise to thiocyanate which binds iodine and makes it unavailable to the body.

Fig. 8.25. A screw-type hand press is used here to express the juice from the grated cassava contained in nylon sacks (Nigeria). (Courtesy I. Nkama.)

Fig. 8.26. Roasting of fermented cassava in the preparation of gari (Ghana).

Fig. 8.27. The fermented cassava is heated under stirring in the trough in the left of the picture. The burning firewood can be seen in the first of the four trollies (Nigeria). (Courtesy I. Nkama.)

Sweet potato

This is not related to the potato but both have tubers. Sweet potato grows in wet tropical regions as a secondary crop. The elongated or round tubers are white, red or purple in colour. The flesh is white or yellow. If yellow, it is rich in vitamin A. The tuber contains starch, some proteins and a little sugar which causes the sweet taste. It contains less moisture than potato (approximately 70% rather than 75%). It may be eaten boiled and mashed.

Taro

This is found in tropical, moist and swampy soil. It is referred to in the Pacific as 'taro', in the West Indies as 'eddo' and in West Africa as 'old cocoyam'. The edible part is the thickened base of the stem. Nutritionally it is of secondary importance.

9

Non-alcoholic beverages

The most important component of beverages is water and that is one of man's most important needs. However beverages contain other ingredients which may make them desirable or perhaps undesirable. Fruit juices generally contain vitamin C and sugar, coffee and tea a stimulant, cocoa and some soft drinks a significant amount of energy. Coffee and tea have little food value but may serve as a vehicle for sugar and milk. Alcoholic drinks will be considered in Chapter 14.

Coffee

About one third of the world's population drinks coffee. A great deal of it comes from Brazil but there are large plantations in India, Indonesia and Africa. The coffee trees are raised from seed in nurseries and planted out after two years. In another two years the trees bear their first berries (Fig. 9.1). Green at first, their skins become red and they are the size of small cherries. Coffee can be processed wet or dry. In the wet process the berries are collected in baskets and sacks and taken to the factory which is always by some watercourse. Here the berries are inspected and slit open. Skin and flesh are sent away on a stream of water while the two seeds inside, the coffee beans, are fermented for 72 h in large tanks of water. After that the beans are washed, and flushed out on to trays where they will dry for about three weeks. At night the trays are covered with plastic sheets to keep out the dew. The skin and flesh are recycled as fertiliser. When the beans are dry their outer skin is removed, they are graded and sorted by colour before they are sent, still green, to the world's big auction houses. They will finally be roasted just before sale.

In the alternative dry process, the berries are simply dried in the sun or a mechanical drier, and the hard shrivelled husk is then stripped off (Fig. 9.1a).

Fig. 9.1. A coffee tree showing the cherry-like fruit and its pickers.

One cup of freshly prepared coffee may contain between 50 and 150 mg of caffeine and 1 mg of nicotinic acid. Caffeine decreases fatigue, increases the amount of physical work that can be done and affects the circulation. If taken in excess it can cause sleeplessness and diarrhoea. Soluble coffee is prepared by drying the liquid infusion either by spray, freeze- or vacuum-drying. Decaffeinated coffee is prepared by the extraction of the green beans with organic solvents or aqueous acids or alkalis. The dried beans are then roasted in the usual way.

Fig. 9.1a. Picking out defective coffee beans in the grading factory in Tanzania. (From Brama, E. *Tea and Coffee*. Hutchinson, 1972.)

Tea

This is second to water the world' s most popular beverage. The country of origin is China where tea was consumed as early as 3000 B.C. In 800 A.D. tea reached Japan via Korea. The first consignment arrived in England in 1657, apparently the first to reach Europe. The Dutch grew tea in Java and later the English in India and Sri Lanka. Tea trees are tall, eventually growing up to 10 m. Harvesting usually takes place by hand-picking the small shoots (Fig. 9.2), but there is also combine harvesting equipment in extensive use. In the manufacture of tea, the young shoots with their leaves are picked and briefly fermented and dried (Figs. 9.3, 9.3a, 9.3b). The material is then broken up and sieved. About 98% of world production is fermented black tea. Green tea is unfermented. The leaves are boiled or steamed to inactivate enzymes and the flavour of green tea is similar to that of the fresh green tea leaves. This type of tea is consumed mainly in China, Japan and Taiwan.

Tea blending is a very specialised occupation normally performed by trained tea tasters. The colour of the tea is also considered and often a colorimeter is used in blending.

Fig. 9.2. The picture shows the bud with first and second leaf from which the best teas are made.

Cocoa

Apparently, the cocoa tree originated in the Orinoco Valley in South America, on the Equatorial slopes of the Andes (Fig. 9.5). Linneus gave the tree the name 'Theobroma' meaning 'food of the Gods' (*Theobroma cacao*). Already in ancient times, cocoa was consumed by the Aztec Indians as a beverage. First, the cocoa pods were fermented and sun-dried. The shelled beans were then roasted and ground and vanilla and other spices added. The mixture was then made into a paste with hot

Fig. 9.3. After plucking the tea leaves are spread out and left to wither in a warm and dry atmosphere for about 17–20 h.

water and drunk, or alternatively, it was whipped into a froth and eaten cold. This was given the nam Xocoatl meaning 'bitter water'. From this the modern term Chocolate is derived.

In the sixteenth century, the Spaniards brought cocoa to Europe and sugar was added to sweeten it. By the seventeenth century, cocoa was a well established European drink. In 1828 van Houten of Holland patented a process whereby much of the fat, the cocoa butter, was expressed from the roasted ground beans. So chocolate powder was invented. In 1847 the British firm of Fry & Sons mixed cocoa butter with sugar and

Fig. 9.3a. Modern tea-withering troughs on a tea estate in Africa. (From Bramah, E. *Tea and Coffee*. Hutchinson, 1972.)

Fig. 9.3b. The fermented green tea leaf is fed into firing machines and changed into black tea. It is then spread out to cool. (From Bramah, E. *Tea and Coffee*. Hutchinson, 1972.)

Fig. 9.4. Mate is an infusion drink popular in South America. It contains caffeine and tannins but is not as astringent as tea. It is drunk through a tube called 'bombilla'.

produced eating chocolate. In 1876, Peter of Switzerland added dried milk and made the first milk chocolate.

In modern chocolate manufacture the beans are first removed from the pods (Fig. 9.6) and cured. They are then fermented for 5–7 days in covered baskets or boxes. The temperature rises due to microbial activity and the sugary pulp is converted into alcohol and carbon dioxide. This allows the flavour of the beans to develop and their astringency disappears. Next the beans are dried to a moisture content of about 5% (Fig. 9.7). The shells are now removed and the beans ground. The ground beans are referred to as 'cocoa mass'. Originally drinking chocolate was made by boiling the cocoa mass in water but this was a rather fatty, sickly

Fig. 9.5. The cocoa pods are directly attached to the main trunk of the tree. In this picture they are ready for picking. (Courtesy Unilever plc.)

and indigestible drink. So today, in the manufacture of cocoa powder, some of the cocoa butter is extracted by hydraulc pressure. High-fat cocoa powder contains per 100 g about 25 g fat, 50 g total carbohydrate and 1250 kJ; low-fat powder only 8 g fat, 60 g total carbohydrate and 840 kJ. Both types additionally contain 3–5 g water, 20 g protein, 5 g crude fibre and 5 g ash. Chocolate contains only 1 g water, but 35 g fat, 60 g total carbohydrate and a little fibre, ash and protein per 100 g. Therefore the energy content is very high, of the order of 2100 kJ/100 g.

Fig. 9.6. Cutting open the ripe cocoa pods and extracting the beans.

Fig. 9.7. After fermentation the cocoa beans are dried in the open for about seven days.

Fruit juices

The most important of these is orange juice which may be preserved by canning or freezing. It can also be dried and obtained as a powder. Nutritionally its most important characteristic is a high content of vitamin C and in some parts of the world it is given to infants and old people who may be at risk from scurvy.

Oranges like lemons, grapefruit, tangerines and kumquats belong to the genus *Citrus*. They grow on small trees or shrubs and when ripe the fruit is usually large and conspicuously coloured. In some species the fruit is predominantly sweet, in others it is acid due to high concentrations of mainly citric acid.

Oranges are grown on a very large scale in Israel, the southern United States and in South America. 'Jaffas' are large, juicy, seedless fruits; 'blood oranges' have flesh with a blood-red tint. 'Seville' or 'bitter oranges' are too bitter to eat and are used predominantly in the United Kingdom for the manufacture of marmalade. The fruit yields about 45% by weight of juice. The fresh juice contains about 50 mg/100 g of vitamin C, the commercial juice about half that. The extracted pulp may be dried and used as animal feed or for the recovery of pectin which is useful in industry as a thickening agent, for instance in the manufacture of some jams. The skin contains essential oils which may be used in food manufacture or perfumery.

Two important acid citrus fruits are the lime and the lemon. Because of the relatively low sugar and high citric acid content the fruit is usually too sour to be eaten but is used extensively as a food adjunct and for its juice.

The fruit of the lime is relatively small, about 4 cm and yellowish-green when ripe. It is a typical plant of the tropics because it is the least hardy of the citrus plants: it is destroyed by frost. Limes are not exported to any extent because the fruit soon dries up and becomes unattractive. A major exporter of lime juice is the island of Dominica in the West Indies. In subtropical parts of the world the place of the lime is taken by the lemon.

Carbonated drinks

Carbonated soft drinks usually consist of water, 1–3% carbon dioxide, 3–10% liquid sugar acidified to pH 3.5, colour and various flavouring additives. These may by synthetic or extracts of herbs, roots, bark, fruits or flowers. The water is carefully purified. If it is too hard it is passed through a water softener and if contaminated it is filtered through carbon

Table 9.1 *Equal sourness of various acids (g/l)*

Ascorbic	3.00	Phosphoric	0.85
Citric	1.28	Tartaric	0.95
Lactic	1.60		

Table 9.2 *Composition of four common soft drinks (per 100 g)*

	Water (g)	Energy (kJ)	Protein (g)	Fat (g)	Total carbo-hydrate (g)	Ash (g)
Coca cola	90	170	Tr	0	10	Tr
Lemonade	95	90	Tr	0	6	Tr
Orange juice	89	140	Tr	0	9	Tr
Tomato juice	93	70	1	0	3	Tr

Tr = trace

or treated with ozone. For this reason bottled drinks are usually safe even if other water is likely to be contaminated. (For visitors not familiar with the tropics boiled drinks like coffee or tea and bottled beverages like soft drinks or beer are often preferred. These may also be used for cleaning one's teeth and ice in the drinks is often avoided because it might be made from contaminated water.)

The carbon dioxide must be free from air or the liquid may 'boil' violently when the bottle is opened.

The most common acid used is citric acid which is naturally present at levels of 6–8% in the juice of unripe lemons. Table 9.1 gives values for equal sourness of the various acids. Very small amounts of benzoic acid are also often used as a preservative.

All beverages with the exception of wine and certain distilled liquors deteriorate with age. The rate of deterioration depends on composition, container, temperature, light and handling. Spoilage is detectable by change in flavour, cloudiness (haze), sediment or floating particles. Beverages are often pasteurised to improve keeping properties. Table 9.2 gives the composition of four common soft drinks.

10

Sugar and other sweeteners

Sucrose

All natural sweeteners except honey are isolated from plant juices either by extraction or crystallisation. The most important sweetener is sucrose. This is obtained either from sugar cane, a member of the grass family growing in the wet tropics or from sugar beet, a root vegetable of cooler areas. Sugar cane provides about 60% of the world's requirements and the most important areas in which it grows are Brazil, India, Cuba, the West Indies, Hawaii and East and West Africa (Fig. 10.1).

Sugar beet is grown mainly in Europe, North America and Canada. Like sugar cane, beet sugar is extracted and recrystallised from water except that from cane, sugar is extracted by crushing rollers and from beet by 'diffusion' i.e. shredding followed by leaching with hot water (see Chapter 13).

When purified, cane sugar is indistinguishable from beet sugar. The concentration of refined sugar is about 99.9% making it the most widely used pure chemical in the world. It has a high energy value of about 1670 kJ per 100 g but contains few other nutrients.

Apart from its sweetness sucrose gives aqueous solutions a high viscosity, has preservative properties and forms 'glasses'. A liquid or supercooled liquid is said to be in the 'glassy state' when it is brittle and its viscosity is so high that it supports it own weight. Some higher alcohols, fats and very concentrated sugar solutions form glasses.

Boiled sweets are in the glassy state and quite clear. In their manufacture a saturated aqueous solution of sucrose is boiled. To avoid crystallisation at high temperature potassium hydrogen tartrate may be added to invert about one-third of the sucrose. As the water evaporates, the temperature of the solution rises. When a sample is removed and cooled to room temperature, it exhibits very typical physical properties, and

Fig 10.1. Before harvest the leaves of the cane can be burnt. That reduces vermin and the burnt cane is easier to cut. However the sugar quickly inverts and burnt cane must be processed within 48 h.

according to these bears the name given by the sugar boiler. At a boiling point of 110°C the thread degree is reached: the sugar is dense enough to form threads when a fork is inserted into it and pulled away. 115°C is referred to as feather degree: the sugar will blow off the fork as feather-like fluff. At the ball degree the sugar solution, once cooled to room temperature, allows a soft ball to be formed between the fingers. Crack degree indicates that the ball is now hard and brittle. Table 10.1 gives the relation between the technological description, boiling point and total solids content of sugar.

Boiled sweets are at the 'hard crack' stage. Unfortunately they are also

Table 10.1 *Relation between technological description, boiling point (B.P.) and total solids content (T.S.) of sucrose solutions*

Technological description	B.P. (°C)	T.S. (%)
Thread	110–113	75–80
Feather	115–119	87–88
Soft ball	121–124	90–91
Hard ball	127–130	92–93
Soft Crack	135–138	95
Hard crack	144–149	97–98

Table 10.2 *Relative sweetness of various sweetening agents*

Sucrose	1.00	Glycyrhizin	100
Glucose	0.5	Aspartame	180
Laevulose	1.7	Dulcin	200
Lactose	0.2	Saccharin	300
Maltose	0.3	Stevioside	300
Cyclamate	30	Thaumatin	3000
Chloroform	40	P-4000*	4000

* 5-nitro-2-*n*-propoxyaniline

hygroscopic and when placed into a moist atmosphere will absorb water. The sweet will then crystallise from its glassy state and as a result becomes opaque and sticky. This development of stickiness is typical during the crystallisation of glassy sugars. It may cause lumping and caking in milk powder and freeze-dried products. The crystals themselves may cause grittiness in ice cream.

Commercial caramel is prepared by heating sucrose to 220°C for 10–15 min. On cooling it sets to a brittle solid with a persistent bitter taste. It is produced industrially to definite colour specifications and is used in colouring beer, vinegar, gravy, coffee-flavoured products and many other foods.

Sucrose is also regarded as the standard for sweetness and Table 10.2 gives a number of relative sweetnesses of various materials in relation to sucrose.

Invert sugar

This is a useful sweetening agent obtained by acid hydrolysis or by the digestion of sucrose with the enzyme invertase. It is a mixture of dextrose and fructose and is often used in industry as an additive to retard crystallisation of sugars in other foodstuffs. It is also extensively used by brewers who add it to the wort (p. 237).

Glucose

Probably one of the most abundant organic chemicals in nature, it was first isolated in the crystalline state by Kirchoff in 1811 from acid-hydrolysed starch. Modern methods make use of enzymic hydrolysis which gives rise to a mixture of dextrose, maltose and several higher saccharides. Mainly manufactured from maize, it is much less sweet than sucrose and is used as an additive in food manufacture. Commercial liquid glucose contains between 12 and 25% water.

Jam

Fruit can be preserved by boiling it with sucrose and allowing it to cool and set in sterile jars. A set is only obtained if there is sufficient pectin which acts as a setting agent. Citric and tartaric acids may be used to liberate the pectin and improve the flavour. These acids also invert some of the sucrose and prevent the jam from crystallising. The fruit should be just ripe, sound and dry. Some fruit e.g. lemon, lime or bitter orange contain satisfactory levels of pectin. Others e.g. pineapple or sweet orange contain too little and commercially prepared pectin or high-pectin fruit or fruit juice must be added. Ideally there should be 5 kg of jam for every 3 kg of sucrose used. After boiling in a heavy pan, the mixture is poured into sterile jars and sealed using waxed circles and cellophane covers or screw caps to prevent entry of micro-organisms.

Honey

Wild honey is probably the oldest sweetener (Fig. 10.2). Bees collect nectar from flowers and add the enzyme invertase. This splits the sucrose of the nectar into dextrose and fructose. The bees then transfer the material to the honeycombs where it loses water. While the composition of nectar is approximately 20% sucrose and 80% water, honey contains approximately 35% dextrose, 40% fructose, 15% water and 10% miscellaneous material containing approximately 2% sucrose as well as proteins, dextrans, organic acids, essential oils, vitamins, minerals, pollen grains, yeasts and bacteria. Honey of certain districts may have the typical aroma of the local flora and heather and orange blossom honey are

Fig. 10.2. Over 7000 years ago the ancient Egyptians already cultivated honey bees. Honey, they thought prolonged life and acted as an aphrodisiac. There is no scientific evidence for this.

easily identified. It is well known that some honeys may contain poisons (mellitoxins). This is very rare but honey from a species of rhododendron (*Azalea pontica*) which grows by the Black Sea has been known to cause disorientation, headache and vomiting. Table 10.3 gives the analysis of various sweetening agents.

Table 10.3 *Analysis of various sweets and sweeteners (per 100 g)*

Material	Water (g)	Energy (kJ)	Total carbohydrate (g)	Ash (g)
Beet or cane sugar				
Brown	2.0	1560	96.4	1.5
White	0.5	1610	99.5	Tr
Dextrose (cryst.)	9.0	1400	91.0	Tr
Maple sugar	8.0	1455	90.0	—
Honey	17.0	1270	82.3	0.2
Sugar cane juice	80	300	17	—
Jaggery	3	1560	97	0.5
Turkish delight	16	1260	80	Tr
Jam	30	1100	70	Tr
Marzipan*	10	1860	50	1.0
Toffee**	5	1800	70	0.5
Chocolate, plain***	Tr	2100	60	1.0

Note: Tr = trace, * 25 g fat, ** 17 g fat, *** 35 g fat.

Jaggery

This is the crude brown sugar obtained from sugar cane or the juice of various palms, notably the sugar palm which grows wild in Malaysia and Indonesia. The flowering shoot is cut off and a vessel is tied to the stump to catch the juice. Up to 1.5 l per day can be obtained for about six weeks if repeated slices are cut from the stump. Jaggery (the Hindi word is goor) is made by boiling the sap in open pans. (Fig. 13.16). The scum is regularly removed and evaporation turns the sap into a thick syrup. On cooling either a semisolid is obtained or if evaporation has progressed sufficiently, solid brown toffee-like sugar cakes can be made.

If the palm juice is left to ferment it becomes an intoxicating drink called toddy, and this distilled is arrack.

Synthetic and unusual sweeteners

The search for new sweeteners is difficult because the relationship between molecular structure and sweetness is not understood. For this reason many of the synthetic sweeteners were found by accident. The first was saccharin discovered in 1879. It was first used by diabetics who were unable to eat food sweetened with sugar. It is now used in many foods although it is prohibited in some countries because of a possible health risk. Its drawback is a distinct aftertaste detected by a few people at the equivalent of 5% sucrose solution. It is about 300 times as sweet as sucrose but relative sweetness is difficult to measure and therefore the values in the literature may vary greatly.

Cyclamate, like saccharin, was also discovered by accident. In 1937, a research student placed his cigarette on the bench on which he had spilled a trace of his organic preparation and found that the cigarette tasted very sweet. The use of cyclamates has recently been prohibited in many parts of the world because cyclamates caused abnormalities in chick embryos and given in extremely high doses, caused bladder cancer in rats. That research has not been confirmed but the ban in the United States and Britain has never been lifted. Cyclamate is about 30 times as sweet as sucrose and has no aftertaste.

Even more recently another sweetening agent was discovered by accident and has been marketed since 1981. It is called aspartame and is a dipeptide of the amino acids aspartic acid and phenylalanine. Aspartic acid alone tastes flat and phenylalanine bitter; combined they are 180 times as sweet as sucrose and possess no aftertaste. Aspartame is metabolised in the body in the same way as protein but it must not be given to the one child in 15 000 who suffers from a rare condition called phenylketonuria. This is a genetic deficiency which makes the child react badly to an intake of phenylalanine.

Problems with many synthetics are aftertaste, off-tastes or taste lag as well as insufficient stability to heat or pH. The most important problem with all synthetic food is possible chronic toxicity which may be very difficult to evaluate. Firstly testing takes a great deal of time and money and secondly rats, guinea-pigs and monkeys which are often used as test animals may not react in the same way as man. People, of course, cannot be used for dangerous tests.

At present there is also an intensive search for natural sweetening agents. Some are intensely sweet and if they have no side effects can be most useful. Thaumatin consisting of two proteins 3000 times as sweet as sucrose has been obtained from the sweet-fruited forest plant *Thaumatococcus daniellii* (Katemfe) which is widely distributed in West Africa. It can for instance be found near Coomasie and Onitsha. Monellin, also about 3000 times as sweet as sucrose, comes from *Dioscoreophyllum commensii* (the serendipity berry which also grows widely in West Africa). Stevioside, as sweet as saccharin, has been obtained from *Stevia rebaudiana* a plant common in many parts of the tropics. Other sweeteners have been isolated from liquorice root (glycyrrhizin) and orange peel.

Two other strange substances have been described. Maltol and its derivative ethyl maltol have little taste themselves but enhance other sweet or fruity tastes. Miraculin is a protein obtained from the juice of the miracle berry *Richardella dulcifera* found in tropical areas of the Pacific and in West Africa. It causes sour substances such as lime or rhubarb to taste sweet within one minute after it has been applied to the tongue. This conversion effect lasts several hours.

11

Foods of animal origin

Unlike plants, animals have the ability to move about; they possess locomotion. This is possible because animal cells other than bone are flexible and not surrounded by rigid walls of cellulose or lignin. As a result animal tissues do not contain fibre. Water content of metabolic tissue lies between 70 and 80%. There is also no starch but the energy store is either glycogen or fat. Particularly the amount of fat can be very variable and greatly affect tissue composition. Protein on the other hand is not very variable and is of good quality.

Lower animals

Amongst the lower animals commonly eaten are crustaceans (shrimps, crabs, crayfish), insects and molluscs (snails, mussels, squids). The edible sea cucumber (*Holothuria edulis*, Trepang) is found on coral-reefs in pacific waters. Most of it goes to China.

The composition of crustaceans is fairly constant. Water content is about 80% and the energy value 335–375 kJ/100 g. The content of fat, carbohydrate and ash is below 2%. Protein varies from 15 to 20% and is similar in quality to that of meat and fish. The PER is of the order of 2.2.

Insect eating is not very common but has been practised since ancient times. Insects can provide extremely useful sources of nutrients and if edible and available in quantity their use must be encouraged. Species of *Cicada* were eaten by the ancient Greeks while various caterpillars were popular amongst the Romans. In more recent times the larva of the coffee-boring moth (*Zeuzera*) is regarded as delicious when eaten with rice and salt by people in Indo-China and Thailand. Flying ants cooked in butter are eaten in many parts of Africa and properly cooked locusts (*Nomadacris*) and lake fly (*Chaoborus*) can be an attractive and nourishing food. Water content is about 60–70% and protein varies from 20 to

30%. As with meat, fat content varies greatly. Values from 2 to 30% fat have been quoted. Ash content varies from 3 to 5% and insects tend to be rich in potassium and phosphorus. Some edible species also contain significant quantities of riboflavin, nicofinic acid, vitamin A and iron.

The molluscs of interest to the food scientist fall into three groups: the gastropods, the bivalves, and the cephalopods. The gastropods include the snails and slugs of which there are about 40 000 species or 80% of all living molluscs. They are found in most parts of the world on land, in the sea or in fresh water. The largest and most striking examples are tropical. The European edible snail was in the past a popular food all over Europe but today it is an expensive luxury. It has been virtually wiped out as a pest of vineyards and a parasite vector, but it is still extensively farmed.

The giant African snail is a major pest in tropical areas. In West Africa various species of *Achatina* and *Archachatina* are eaten extensively. In Ghana it is the largest single item of animal protein. It is collected in the rainy season because it is difficult to find in the dry season. The snails are very large and grow up to one foot in length. They are boiled or fried, eaten with fufu, or added to soups or stews. They are also useful as snacks. They are preserved by removing them from the shell, eviscerating and smoking them. They are finally sun-dried. The recent spread of *Achatina*, although highly undesirable from the agricultural point of view, made it available as part of the diet in many areas of Africa, Malaysia and Japan. On occasions, snails have been reported to be toxic. This does not appear to be true although insufficiently cooked snail which has fed on contaminated or infectious material could obviously be a threat to health. Also some people are allergic to snails. Of the marine gastropods, cockles and abalones (Fig. 11.1) are of some importance as food.

The bivalves are the second largest class of molluscs containing approximately 7500 species. All are aqueous – there are no land forms. There is extensive farming of oysters. Bivalves are usually eaten grilled, pickled or boiled. Occasionally, mussels are infested by the dinoflagellate *Gonyaulax catenella* which causes severe food poisoning.

The cephalopods present the highest evolution of invertebrate development. There are only approximately 600 species. The animals are free swimming with large eyes and a ring of tentacles bearing suckers. Examples are octopus, squid and cuttlefish. The animals are eaten extensively as food in tropical and subtropical countries, particularly Japan. They are fished with nets like other fish and eaten boiled, canned in oil or fried.

The composition of the molluscs is fairly uniform. Water content is

Fig. 11.1. Abalones are found in the coastal waters of the East and West Pacific. This picture is by the Japanese artist Utamaro Kitagawa (1753–1806).

about 80%, energy 300–400 kJ/100 g, fat 1–2%, total carbohydrate 2–4% and ash 1–2%. Protein content varies somewhat and is often inversely proportional to water content. The oyster with 85% water contains 8% protein, the giant African snail with 80% water, 10% protein and the abalone with 75% water, 18% protein.

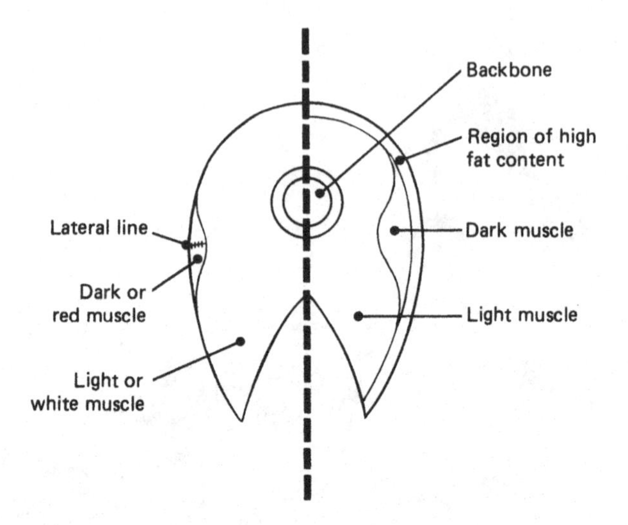

Fig. 11.2. Diagrammatic representation of a section of white fish (left) and fatty fish (right).

Fish

Fishing is as old as man and fish have always been an important source of food. The most important variable in composition is the fat content and therefore fish are divided into white fish and fatty fish. The ratio between the highest and the lowest fat contents is more than 300 : 1. Figure 11.2 shows a diagrammatic representation of a white fish (left-hand side) and a fatty fish (right-hand side).

Fat content varies very much not only with the type of fish but also with season. Therefore the analyses reported in the various food composition tables are only very approximate. Fatty or 'dark' fish (and that group includes salmon, sardines, pacific mackerel, bream and bonga) contain about 65–70% water, 16–20% protein, 10–20% fat, 1–2% ash and 835–1050 kJ/100 g. The dark colour is due to the oil between the muscle fibres.

White or 'lean' fish stores its fat in the liver. This group includes cod, haddock, barracouda, bambangin, parang, puffer and grouper, and these contain about 80% water, 16–18% protein, 0.3% fat, 1–1.5% ash and 335 kJ/100 g. The PER for both types of fish is about 2.3 and the total carbohydrate content is usually below 1%.

There are three other fish which should be mentioned. The first is the tuna fish or 'tunny' (Fig. 11.3). It is a large fatty fish of tropical and subtropical waters. The second is the anchovy. This was extremely important in Peru as a source of fish meal until overfishing in the 1970s caused the closure of dozens of small harbours and factories all along the

Fig. 11.3. Tunas are found in the warmer parts of all oceans. It is one of the largest sea fishes known and may reach a length of 3 m and a weight of 500 kg.

coast. The third is stockfish. This is a commercial term for salted and dried fish such as cod, hake, haddock or ling. It is exported from Canada and Scandinavia to many parts of the tropics (Fig. 11.3a).

Some fish provide fat which can be separated and from others the intestine and liver are used as food. If the latter is eaten the gall bladder must be carefuly removed.

The most important fish offal eaten is the roe (fish eggs). At about 70% water it contains 20–25% protein and is rich in nucleic acids. Fish oils usually obtained from the liver are good sources of the fat-soluble vitamins, and small fish which are eaten whole are important sources of calcium.

In manufactured fish products, the type of fish is usually readily analysed by means of electrophoresis. With this technique the proteins are separated in a gel by an electric field, and after staining show lines in the gel which are typical of the type of fish (Fig 11.4).

Meat

Meat is the flesh of animals used as food. The great majority of people the world over obtain a large proportion of their nutrients from meat. Meat is

Fig. 11.3a. Dried fish ready for sale in Hong Kong. (Courtesy F. A. Leeming.)

easily digested and absorbed. It also has a high satiety value but it is expensive. Generally the higher the standard of living the greater is the consumption of meat. Nutritionally, any meat that is healthy is good to eat, whether it be frogs, dogs, guinea-pigs, rats or bush meat.

Livestock in the tropics is often of very much poorer quality than that in the developed world. For instance livestock zones in Africa are largely determined by the availability of feed and water and that of cattle by prevalence of the tse-tse fly. Immunity of local humpless cattle (Fig. 11.5a, b) often runs parallel with small size and debility. The humped

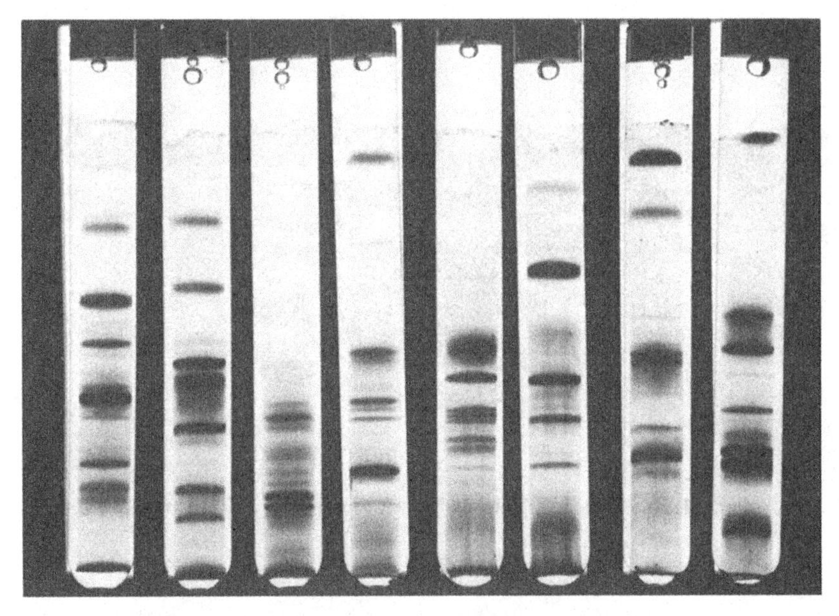

Fig. 11.4. Separation patterns of raw fish (left to right: megrin, witch, lemon sole, plaice, turbot, salmon, haddock, cod). (Courtesy I. M. Mackie: Crown copyright by permission of the Torry Research Station.)

Fig. 11.5a. The water buffalo is common throughout Asia. A placid and contented animal it is an excellent swimmer. Its meat has a somewhat musky flavour. (Courtesy H. J. S. Taylor)

Fig. 11.5b. Water buffalo in India ploughing a field of paddy-rice.

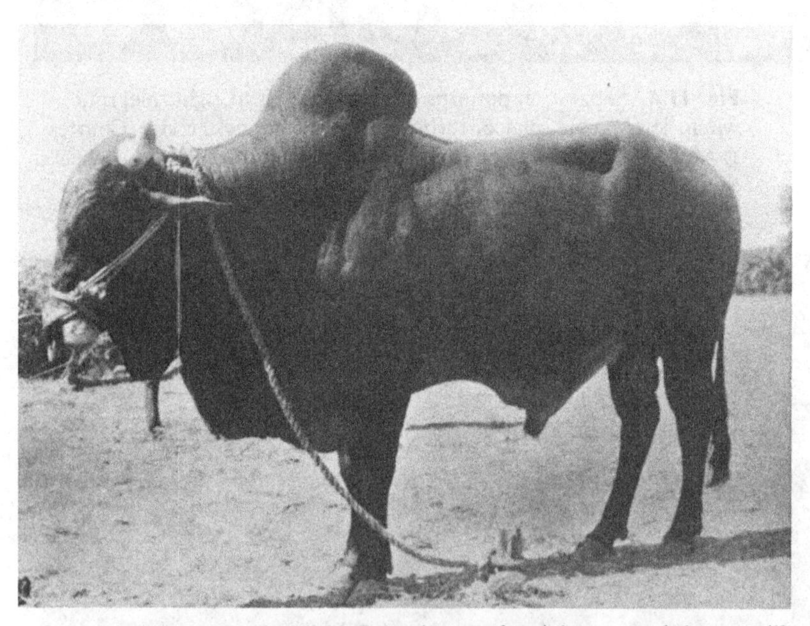

Fig. 11.6. The humped Zebu (*Bos indicus*) has a much lower milk yield than European cattle (*Bos taurus*) but is more resistant to heat. The picture above shows a Zebu bull.

Zebu cattle (Fig 11.6) have a much lower milk yield than European cattle but are more resistant to heat. Sheep are often small, short-legged and hairy. Goats (Fig. 11.7a, b) are common because they withstand the heat better than sheep and are less demanding. One can see a goat crunch

Fig. 11.7a. In many parts of the tropics goats provide more meat than any other animal, particularly in Africa and Asia. (Courtesy T.R. Hall.)

Fig. 11.7b. Sheep were introduced into Australia shortly before the beginning of the 19th century. Since then the stock has been carefully selected and improved.

Fig. 11.7c. As a beast of burden the llama of Peru has yielded to the mule but it is still highly prized for its wool. The meat is agreeable and resembles mutton. (Courtesy H. J. S. Taylor.)

dried up leaves as we would eat cornflakes. Goat meat is consumed by Africans in greater quantity than any other.

Chickens are ubiquitous in the villages but they are often very small and scrawny and produce only about 100 usually tiny eggs per year and these are often stale when found.

Meat consists of approximately 70% water, 20% protein, 1% carbohydrate and 1% ash. As for fish, the fat content varies widely between 1 and 20%, and greatly affects both composition and minor components. For this reason, all analytical data on meat are only approximate guides. There are variations in nutrients of 15% and more. Table 11.1 shows the difference between lean and fat beef.

The internal organs of the animal (the offal) do not differ greatly in fat content, but that of the tongue may vary from 10 to 23%. Table 11.2 gives the composition of beef offal.

The biological value of meat is high but not as high as that of milk and eggs. The PER is 2.3. Nutritionally, the amounts of the fat-soluble vitamins A, D, E and K are negligible. The content of calcium is very low but meat is high in uric acid precursors. Iron content is fairly high and about 30% of the iron is available. Cow's coagulated blood contains about 40 mg/100 g of iron.

Meat is an important source of the B vitamins. These are characteristi-

Table 11.1 *Proximate analysis of lean and fat beef muscle (per 100 g)*

	Water (g)	Energy (kJ)	Protein (g)	Fat (g)
100% lean beef	72	630	22	6
67% lean, 33% fat beef	58	1212	18	23

Table 11.2 *Composition of beef offal (per 100 g)*

	Water (g)	Energy (kJ)	Protein (g)	Fat (g)	Total carbo-hydrate (g)	Ash (g)
Heart	78	460	17	4	0.7	1
Kidney	76	543	15	7	0.9	1
Liver	70	585	20	4	5.0	1
Lungs	79	418	18	2	0	1
Tongue	68	836	16	15	0.4	1
Brain	79	523	10	9	0.8	1.5
Blood (coagulated)	78	380	21	0	0.5	0.7

cally heat labile and water soluble. Considerable variation in vitamin B content exists not only from one species to another but also between different muscles of the same animal. Differences of 15% within the same animal are not unknown. Since the B vitamins are water-soluble, percentage content decreases with increasing fat content of the tissue. B vitamin content, as indeed meat composition in general, is affected by type of feed, season, sex and age of the animal. Meat is a good source of niacin (4–12 mg/100 g) and riboflavin (0.1–0.27 mg/100 g). Liver contains about six times the riboflavin content of other meat or offal. Thiamin content of most meats is about 0.1–0.3 mg/100 g but pig meat is particularly high in thiamin, containing up to 1 mg/100 g.

Meat structure

Meat other than offal is generally muscle (Fig. 11.8). It is surrounded by connective tissue which terminates at both ends in tendons fixed to bones. The muscle may be overlayed with pieces of fat. Within the muscle one finds a large number of muscle fibres packed into bundles by more connective tissue which itself may contain blood vessels, nerves and more

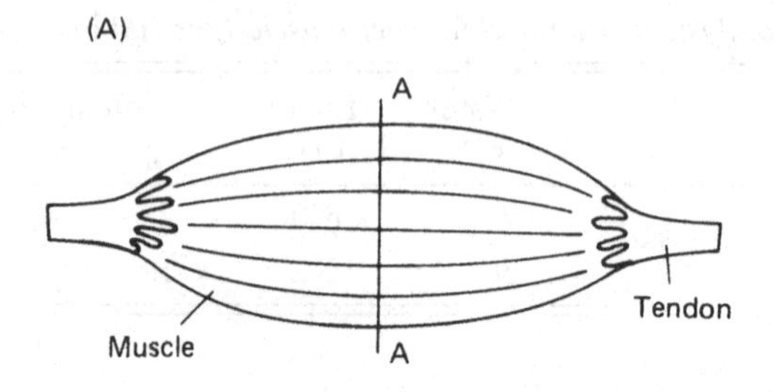

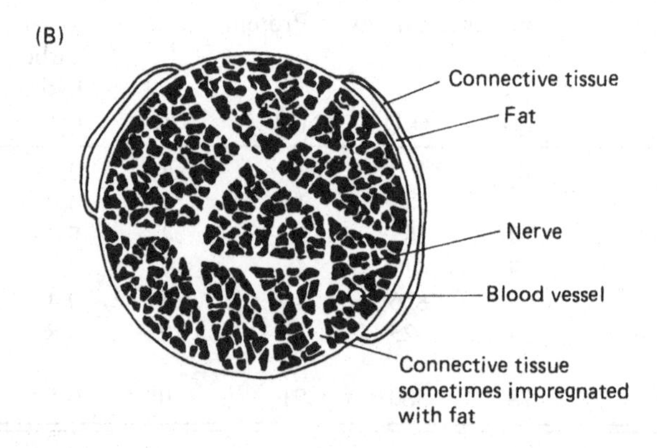

Fig. 11.8. (A), a muscle showing tendons and section at AA; (B), a transverse section of a muscle at section AA.

fat cells which give rise to marbling. The size of the muscle fibre bundles affects the texture of the meat, large bundles giving a coarse texture. The muscle fibres themselves are the contractile elements of the muscle.

Post-mortem changes

While the animal is alive, tissue enzyme reactions are reversible. In death they become irreversible because breathing has stopped and there is no longer a sufficient oxygen supply to the tissues. Microorganisms are also involved in post-mortem changes. Soon after slaughter rigor mortis sets in causing a shortening and hardening of the muscle tissue. After relaxation of rigor (5–20 h depending *int. al.* on temperature), 'ageing' takes place. At first this improves the texture and flavour of meat but later the meat will putrefy due to both tissue enzymes and microorganisms.

Comminuted meat products

Comminution means reducing a substance, in this instance meat, to a small particle size. The term 'meat' is used here in its widest sense including the entire animal except bones, teeth and hair. The products include hamburgers, meat pies and sausages. It seems that all these products are originally of either European or Chinese origin. Apart from finely chopped or liquidised meat, these products contain spices and fillers such as breadcrumbs, rusk (biscuits specially baked and ground for the purpose), cereal or legume flour, milk powder, various starches, pumpkin or marrow. Sausage casings can be obtained from the intestines of sheep, cattle or pigs. Synthetic casings are also available.

Meat inspection

Most developed countries of the world have strict health standards for food. The evaluation of meat requires trained veterinary inspectors who are often not available in sufficient numbers in the developing world. For this reason skilled veterinary teaching facilities must have a high priority in the tropics.

After the animal has been slaughtered, bled, skinned and eviscerated, the carcase should be cleaned and inspected. Apart from parasites such as trichinosis, tapeworm and liver fluke, the meat may contain the organisms of various animal diseases such as anthrax, tuberculosis, brucellosis, erysipelas and foot and mouth disease. Finally, a bacteriological assessment should be made for paratyphoid bacteria, *proteus, coli*, the various cocci, *Clostridium* and *Salmonella*. Some of these, for instance some specis of *Coccus*, *Clostridium* and *Salmonella* present serious health hazards to man. The animal diseases may affect livestock and so indirectly the usefulness of the carcase for food. Others, like the parasites, may be hazardous to man if the meat is not properly cooked. All meat must be thoroughly cooked and that includes the centre.

Eggs

The eggs of many species of birds are eaten in various parts of the world. They include the eggs of the Guinea hen, the domestic hen, duck, turkey and many others. Both the structure and composition of eggs are very similar. A transverse section of the hen's egg is given in Fig. 11.9.

The calcareous shell surrounds the two shell membranes and underneath these lies the egg white or albumen. This consists of three layers,

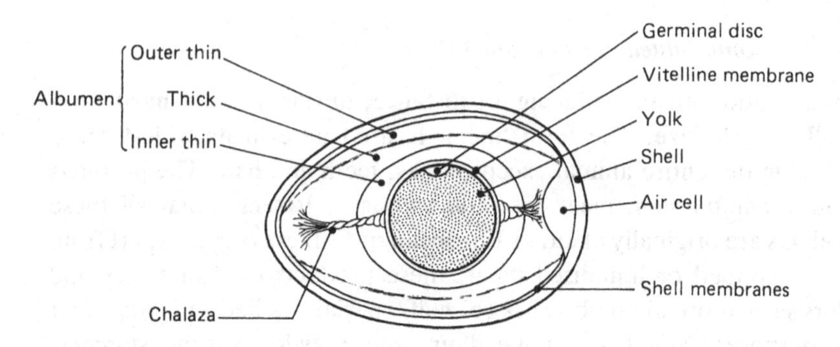

Fig. 11.9. Section of the hen's egg.

the outer thin (23%), the thick (57%) and the inner thin (20%); 85–90% of the white is water, the rest mainly protein. The central yolk makes up about 50% of the total solids consisting of about 16% protein and 32–36% lipids (including 5% cholesterol).

Nutritionally, eggs are a very valuable, if rather expensive source of nutrients. Raw eggs contain about 75% water, 670–795 kJ/100 g, 13–14% protein, 10–15% fat, 1–2% total carbohydrate and 1% ash. In an assessment of the nutritive properties of protein the egg protein almost serves as a reference standard. The PER is 3.92. Apart from being an important source of unsaturated fatty acids, mainly oleic, eggs contain iron, phosphorus and other trace elements. Vitamins A, E, K and those of the B group including vitamin B_{12} are present. The content of vitamin D is second only to that of the fish liver oils. On the other hand, eggs contain only a trace of vitamin C and are low in calcium. The composition of raw and boiled eggs is much the same. As expected, fried and scrambled eggs have a slightly higher fat content, and the latter may also contain milk solids. The fat content of fried eggs is approximately 17%, that of scrambled 15%. Heated egg even if hard boiled is well absorbed but raw eggs are not.

There are a number of misconceptions about eggs. Shell colour and intensity of yolk colour are not an indication of nutritional value. Also, whether an egg is fertile or infertile has no effect on nutritional value. Similarly, the nutritive value of eggs obtained from battery hens is the same as that obtained from free-range animals provided the former have been properly fed. The age of an egg can be seen from the thinness of the white when broken (Fig. 11.10). It is also possible to shine a light through the unbroken egg. This process is known as 'candling'. If the egg is bright and translucent it is fresh, if it is quite opaque it is too old and spoiled. On

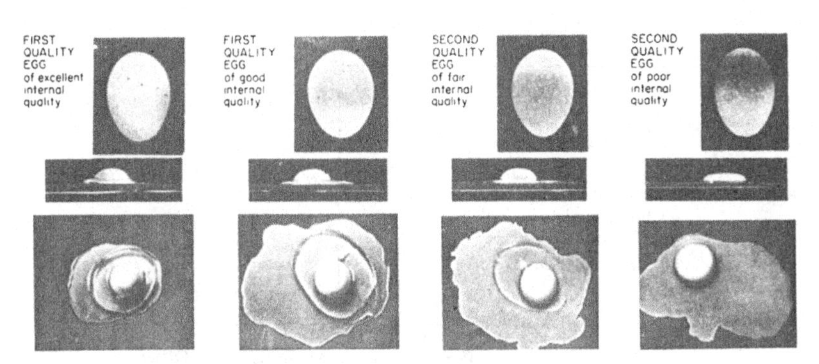

Fig. 11.10. The assessment of the quality of a hen's egg.

storage, foaming and baking as well as organoleptic properties may suffer, but there is little change in nutritive value.

Food poisoning is rarely caused by eggs although *salmonella* has been found in cracked eggs. Some people are known to be allergic to eggs, even in extremely small quantities. They have been known to fall ill after eating a bun, the top of which had been brushed with egg before baking.

For industrial purposes eggs are often processed on a large scale. The most important processes are freezing and dehydration for preservation and de-sugarisation. The removal of the very small amount of sugar is required to retard the deterioration of preserved eggs.

Milk

Milk is produced by the mammary gland of all female mammals. It is the link between mother and young until the latter are able to digest other food. It would appear therefore that milk is the perfect food, but this is not so. Milk is low in iron and, therefore, in some animals including man, iron is stored extensively in the liver of the foetus. Most milks are also deficient in vitamin D. Table 4.6 gives the analysis of human and cow's milk and Fig 11.11 its more important constituents.

The main carbohydrate of milk is lactose sugar. This can be fermented with some bacterial cultures as, for instance, in the preparation of kephir. The resulting fermented milk contains approximately 0.7% alcohol and some lactic acid. Milk is an excellent medium for microorganisms. One member of the tuberculosis bacillus group, *Microbacterium bovis* (*M. tuberculosis*) causes T.B. in both man and cattle. For centuries this most serious disease has been transmitted through milk, although it is caused by one of the few major pathogens which does not itself grow in milk.

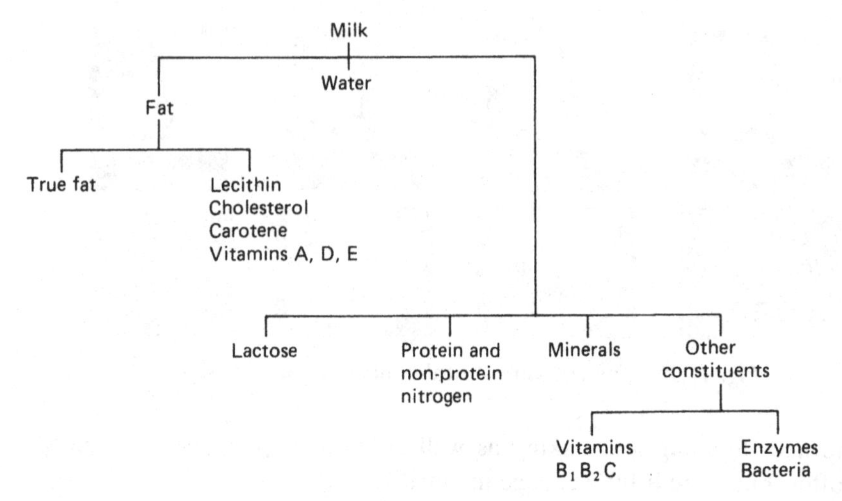

Fig. 11.11. Some of the more important constituents of milk.

Being a non-spore former, the organism is readily killed by boiling or pasteurisation.

The success of milk pasteurisation is checked by the determination of the enzyme catalase which occurs naturally in milk. It is denatured at the same temperature at which *M. bovis* is destroyed. As an additional precaution all tuberculous cattle should be slaughtered and destroyed.

Milk is an emulsion and the fat is found in globules approximately 0.003 mm in diameter (Fig. 2.10). The tendency of the fat globules to cluster is important in the separation of cream either by centrifugal force or by gravity. Table 11.3 gives the composition of milk, thin and thick cream.

The protein of milk has a high PER of 3.09 and the most important of the various proteins is casein. This is precipitated either at pH 4.6 or by the enzyme rennet. (Some plant juices, for instance the leaf extract of the Sodom apple used in West Africa for the purpose, have a similar effect.) As a result the milk 'curdles' and separates into curds and whey. This is the first step in cheese production. Cheese is the natural product of sour milk. Its manufacture has always been a practical way of preserving some of the nutrients of the milk. According to legend, the first cheese was made by an Arab tribesman who carried goat's milk in a bag made from a sheep's stomach. The sun's heat and the rennet from the stomach separated the milk into curds and whey. The curd is the soft cheese which is still eaten in Asia and Eastern Europe today. Basically, cheeses are made by microbial fermentation. First, the milk is coagulated by either rennet or a similar enzyme, or a bacterial starter which produces acid

Table 11.3 *The composition of thin and thick cream and milk (per 100 g)*

Constituent	Milk	Thin cream	Thick cream
Water (g)	87.9	63.9	39.4
Fat (g)	3.5	29.3	56.1
Lactose (g)	4.6	3.5	2.3
Protein (g)	3.5	2.8	1.8
Ash (g)	0.8	0.5	0.4

from the lactose contained in milk. The milk may also be coagulated by the direct addition of acid. The curds are then cut and heated to release the whey. The warm curd is then seeded with the appropriate bacteria or fungi (*Lactobacillus*, *Penicillium*) and the mass moulded. Finally, the curd is fermented sometimes under precisely controlled conditions of temperature and humidity for as long as a year. In processed cheese manufacture several cheeses are mixed and ground together.

Other ways of preserving the food value of milk, while reducing its bulk, are the preparation of butter and milk concentrates. These will be discussed in Chapter 13.

PART THREE

Food preparation and preservation

12

From kitchen to food factory

Basic food processing in the kitchen is really very similar to that in the food factory except that the scale is quite different. In the kitchen a meal for just a few members of the family may be prepared, while in a factory food for thousands may be processed. This large scale of production usually brings with it some specialisation. One factory may concentrate on baking, another on freezing food for convenient storage and still another on canning. Accordingly the tools used, the equipment, are different and often very specialised.

The traditional kitchen

Just as cooking is different in various parts of the world so the kitchen equipment varies. Nevertheless a number of items are usually found.

One might first consider a typical West African kitchen.

On the hearth there may be a pot stand as well as large and small pots made of earthernware, iron or aluminium. There may also be handy bottles for salt, potash, pepper and oil. On the racks are found covered baskets and calabashes for food storage as well as other kitchen implements like plates, ladles, pestle and mortar, graters, sieves and wooden or iron spikes to test whether the food is cooked. Near the floor but raised for easy cleaning may be covered buckets for drinking water and pitchers. Much of the cooking would be done out of doors (Figs. 12.1, 12.2).

On the Indian subcontinent there would be both an indoor kitchen and an outdoor one in the courtyard. During the monsoon and the cooler season cooking is carried out indoors, and on warm winter days and in the summer season outdoors. There is an open hearth and a fuel store for firewood or cow dung cakes. The cooking utensils are usually made of brass or earthenware. They are arranged on shelves or in alcoves in the

Fig. 12.1. In many parts of the tropics much food processing takes place in the open.

Fig. 12.2. Controlling a wood fire is not quite as easy and convenient as controlling a gas or electric cooker.

walls. The dinner service is made of stainless steel or brass and kept in a small cabinet. Spices, particularly curry powder, are kept in well sealed containers. The majority of houses now have a water pump but in the past the only source of drinking water was the village well, a focal point of social activity. The water carrier members of the community are now redundant.

In the traditional Chinese kitchen, small rice bowls and chopsticks would be noticeable. About a third of the world's people eat with knives and forks, another third with chopsticks, and a third with their fingers.

A modern kitchen

The three most important factors in the design of a modern kitchen are efficiency, hygiene and safety. Efficiency because the housewife's job must be made as easy as possible. Hygiene because the health of her family and herself depends largely on cleanliness. Safety because the kitchen with boiling liquids, open flames and possible electrical hazards is the most dangerous place in the home, especially for the young and the elderly.

The ideal modern kitchen is compact with plenty of cupboards and smooth, clean, washable cooking surfaces. Cupboards which are frequently used should be easy to reach, neither too high nor too low. They should open and close easily and tightly. There may be flyproof grids in the doors for aeration. There should be plenty of light and a sink and drainer unit with ample supplies of good quality hot and cold water.

There is an electric or gas cooker sometimes with oven, or a paraffin pressure burner which should be firmly fixed. The kitchen is ergonomically designed with a U or L shape so that the three primary areas, the sink, the cooker and the preparation surface are close together.

The ideal kitchen is fitted with a refrigerator and a washing machine. There are plenty of power points for all electric equipment such as kettle, electric iron or mixer. The walls are often tiled for easy cleaning and the floor has a washable covering such as plastic tiles or stone. There is a washable waste-disposal bin, often foot-operated. Pots and pans are made of iron or aluminium, dishes and plates of glass or porcelain and the cutlery of metal. All these should be easy to clean. Some kitchens have an eating area attached but in some homes kitchen and dining room are separate. Figure 12.3 shows a modern kitchen.

Fig. 12.3. A modern kitchen.

The food factory

Just as in the modern kitchen, the three most important factors in the design of a food factory are efficiency, hygiene and safety. A modern food factory is much larger than a kitchen, so more people are employed, more raw materials must be bought, more wholesome food produced, more waste removed and above all the enterprise must make a profit. That after all is its purpose.

Economics

The site of the factory must consider the source of raw material, water and energy supply, labour and outlet for the finished product. A meat-canning plant is usually found near a slaughterhouse, a vegetable-processing plant near the farms which suply it, a flour mill using imported wheat near the harbour and a fish-processing plant by the sea, river or lake.

Regular supplies of raw materials are most important and often long-term contracts with a producer are required. If such contracts are not made it can happen that the competing wholesaler who supplies the markets can pay the producer more and the supply to the food factory dries up. A food factory must operate at full production for much of the year to be economical and it must not stand empty for months.

Labour is also most important. The workpeople should be intelligent, clean and tidy and should themselves have good standards of living. People living in poor accommodation cannot appreciate the hygienic requirements of a modern food factory. The people should also be satisfied with their working conditions because no enterprise can be efficient with continuous changes of staff. Only recently has there been systematic training for workers in food factories in many developing countries, although the large international companies have practised 'in-house' training for many years. It is, of course, of particular importance that the top management of the factory is properly trained.

Clean water and a satisfactory and continuous energy supply must be available. How this is obtained depends on the factory and on supply facilities. Canneries use a great deal of water, biscuit factories very little. In some areas regular supplies of water and electricity are provided by the authorities; in others it is best if the factory supplies its own needs.

Water supplies in the tropics are often a problem. In the rainy season rivers may become raging torrents and in the dry season there may be hardly any water at all. For most food processing treated water is essential and that often requires the installation of one's own water-treatment plant.

Electricity is also often a problem. Even if the supply is not interrupted altogether there may be voltage fluctuations. When that happens electrical equipment becomes unreliable. Installing one's own generator, just as running one's own water treatment plant, means higher costs. Finally, there must be an outlet for the finished food product. It must be pleasing, wholesome and of the right price in the right place or the food factory will soon have to shut.

Hygiene

The factory must be kept clean and free from dirt, refuse and effluvia. There should be no mould or insect infestation. Floors should be well drained and waterproof if that is necessary, and the walls painted. There should be adequate ventilation, lighting and temperature control. The factory must not be so overcrowded as to be a danger to health. Sufficient lavatories and washing facilities must be provided. The health of the workpeople should be checked regularly to see that they are no danger to each other and that they do not contaminate the food they are processing.

Safety

Moving machinery such as wheel, transmission belts and rotating knives must be well guarded. Pits, vessels or sumps must be fenced to prevent people from falling into them. Moving machinery should not be cleaned whilst operating and lifts and hoists must be safe. Chains, ropes and lifting tackle must be of adequate strength.

Floors, passages and stairs must not be obstructed and there must be safe access to the place of work. There must be no dangerous fumes, smoke, gas or dust which can be a danger to health and pressure vessels such as retorts and boilers must be fitted with safety devices. They must be inspected regularly because corrosion can weaken them in time.

Safe exits must be provided in case of fire and these exits must never be obstructed or locked. Fire-fighting equipment must be available and in good order.

There must also be facilities for wholesome drinking water, washing, resting, first-aid and accommodation for clothing. Working clothes and hats should be provided and cleaned regularly.

In many countries there are legal requirements covering hygiene and safety at work and particularly women and young people working in factories are especially protected.

13

Food preparation

Not all foods can be eaten as they are found and two basic operations are common.

The first is a refining operation and a number of these are considered in this chapter. Important ones are the removal of bran from cereals in the milling process, the isolation of sugar from sugar cane or beet, the removal of protein from plant and animal products and the separation of oil or fat from milk and plant sources.

The second operation is heating. Often food must be boiled or baked to make it edible. The boiling of vegetables, meat and fish and the baking of cereals will be considered.

Refining operations

The refining process is aimed at isolating the edible from the inedible material. With meat, bones and fat are often removed. These contain no water and the fat is high in energy content. Therefore lean meat has a slightly higher water and lower energy content than the meat as bought.

The loss on preparation of vegetables (peel, core, seeds) varies very much with the type and condition of the vegetable, consumer habit and method of preparation. Weight losses vary from 10% for tomato and okra to 30–35% for banana and mango. For pineapple the weight loss can be as high as 50%. Often the outer layers of vegetables may contain more nutrients than the inner ones and so nutrient loss may be significant.

Many vegetables are peeled mechanically but industrially other methods are often employed. For instance, English potatoes can be peeled by abrasion, by boiling in caustic soda solution (lye peeling), a saturated solution of sodium chloride (brine peeling) or by steaming them at 100°C or above (steam peeling). In all instances the skin is easily rubbed or washed off after treatment. Steam peeling is rather expensive but gives a superior product. With lye treatment the removal of the chemicals presents some problems. The examples given below further illustrate various methods of refining.

Cereal processing

The object of cereal milling is to remove the bran and germ from the seed and so free the endosperm. This is left whole as in rice milling, ground into coarse pieces (wheat semolina, maize grits), or milled into flour.

Four topics will be considered briefly: (a) traditional milling, (b) modern dry milling, (c) modern wet milling and (d) rice milling.

(a) Traditional milling

Three types of mill are common, the wooden pestle and mortar mill (Figs 13.1, 13.2), the saddle stone (Fig. 13.3) and the rotary quern (Fig. 13.4). On the Indian subcontinent the latter is referred to as the domestic chakki. There must be thousands of these in the villages.

Figure 13.5 shows the flow diagram of the traditional sorghum milling process in West Africa. The grain is first wetted and placed into the mortar. It is then pounded with the pointed end of the pestle and the bran is loosened by attrition. The seeds are then winnowed and the bran separated. The endosperm is returned to the mortar and ground using the blunt end of the pestle. The fine material is now sieved out. This is the flour. The coarser material is stone-ground on a saddle stone and added to the flour.

These traditional methods are laborious and time-consuming. The procedure described above will yield about 2.5 kg of flour per h. (A medium-sized modern flour mill will produce 2500 kg per h but capital investment is very high.) For this reason attempts have been made to increase production. In India the larger version of the domestic chakki is the kharas. The two mill stones are gravity fed through a hopper and the upper stone, the 'runner', is turned by bullocks, mules or camels. The gharat is the same but driven by a waterwheel. Where electricity is available in the villages the chakkis are power driven and the output is higher. A number of fairly efficient mills have also been designed in the context of intermediate technology. Small hammer mills fitted to bicycles, diesel or electric engines are in use in some villages and hand-operated winnowers to separate bran from endosperm are becoming more common (Fig. 13.6). Separation of germ and bran from the endosperm is not as efficient as in a modern flour mill. The advantage

Fig. 13.1. Traditional pestle and mortar milling. Such mills are found e.g. in Central America, Africa, the Indian subcontinent and South East Asia.

Fig. 13.2. In the far East traditional pestle and mortar mills are often foot operated.

Fig. 13.3. The saddle stone mill consists of a flat stone slab and a cylindrical 'rubber'. (Courtesy E. O. I. Banigo.)

with traditional methods is that some of those nutrients are retained which are discarded with modern methods. This is because some vitamins and minerals are associated with bran and germ. The disadvantage is that these relatively inefficient milling techniques do not remove mycotoxins effectively because, if present, these also are associated with the outer layers of the grain.

Fig. 13.4. This Indian girl uses a rotary quern to grind wheat. (Courtesy A. B. Chaudhri.)

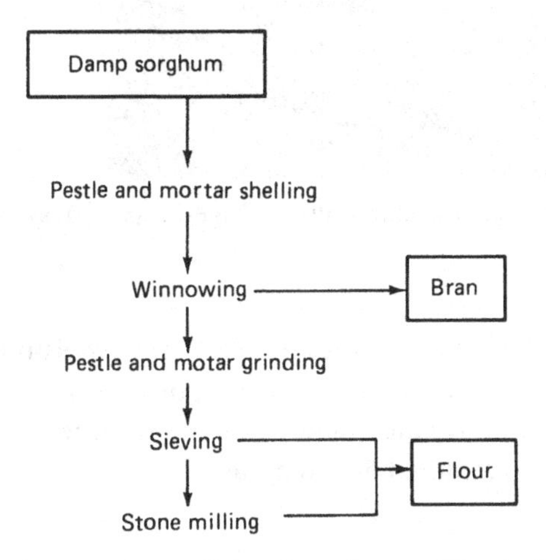

Fig. 13.5. Traditional sorghum milling in West Africa.

(b) Modern dry milling

This is used for maize, wheat and sorghum. Industrial maize milling can serve as an example. The grain is first cleaned carefully by sieving, scouring and washing. It is then tempered by adding water to 20–25% moisture and holding the grain in bins for up to 6 h to equilibrate. It is then passed to an attrition mill which separates hull and germ and leaves

Fig. 13.6. A powered maize mill in use in Nigeria. (Courtesy A. Andah.)

large grits of endosperm. These are then dried to 15–18% moisture and repeatedly milled and sieved until the desired portions have been obtained. Figures 13.7a, b show the roller mills and Figure 13.8 the sifters. Figure 13.9 shows a simplified flow diagram.

(c) Modern wet milling

For maize and sorghum which are also usually milled dry there is an alternative wet-milling process. The object of this is mainly to obtain a purer starch with less contamination by fat, fibre and protein. This starch is either used directly in the food industry or converted by enzymatic or acid hydrolysis into dextrose syrups. A great deal of this is used in the fermentation industry. Figure 13.10 shows a typical wet milling diagram for maize. That for sorghum is similar.

Fig. 13.7a. The roller floor in a modern mill. (Courtesy H. Simon Ltd.)

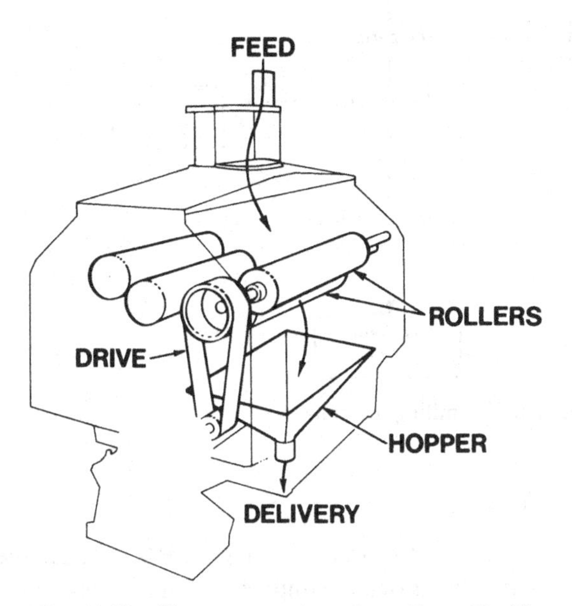

Fig. 13.7b. The construction of a roller mill. (Courtesy H. Simon Ltd.)

Fig. 13.8. Plan sifters in a modern mill. Each section contains 6–8 sieves. (Courtesy H. Simon Ltd.)

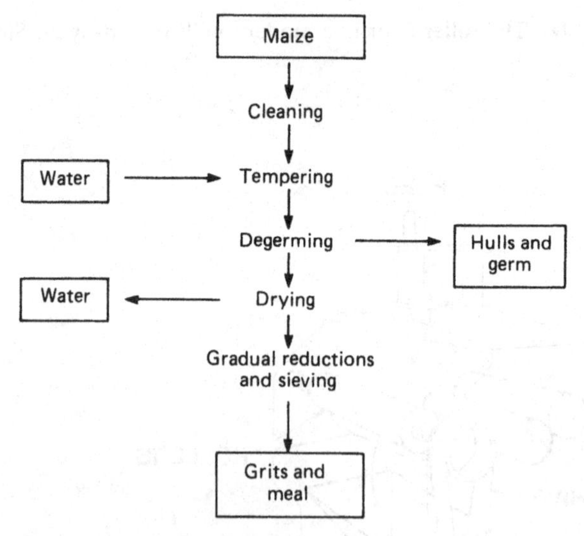

Fig. 13.9. The dry milling of maize.

(d) Rice milling

The milling of rice is different to that of other cereals because the end product is not flour but consists of large unbroken white grains. Rice flour and broken kernels are of lower commercial value. Figure 13.11 shows a diagram of the rice-milling process.

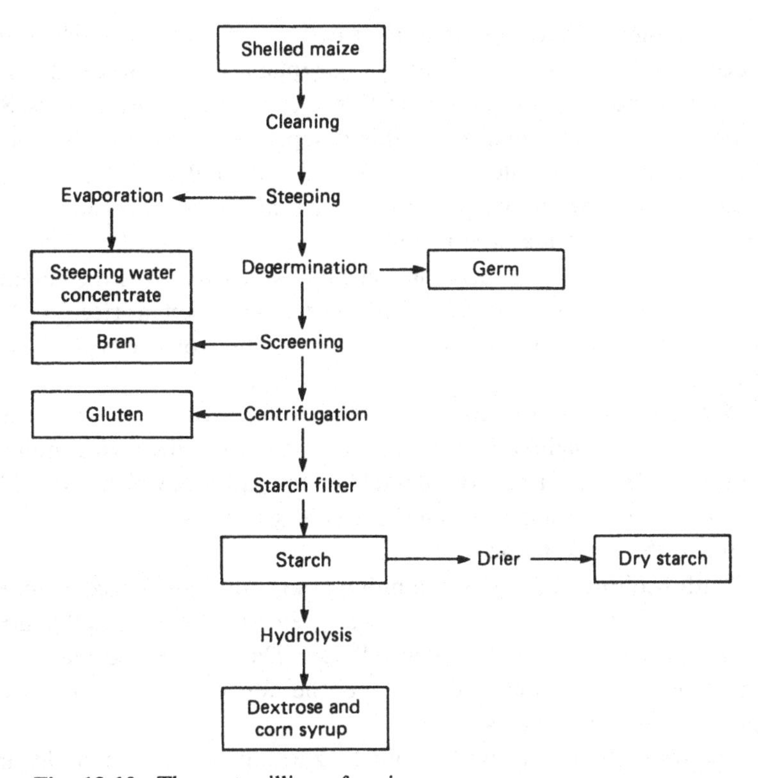

Fig. 13.10. The wet milling of maize.

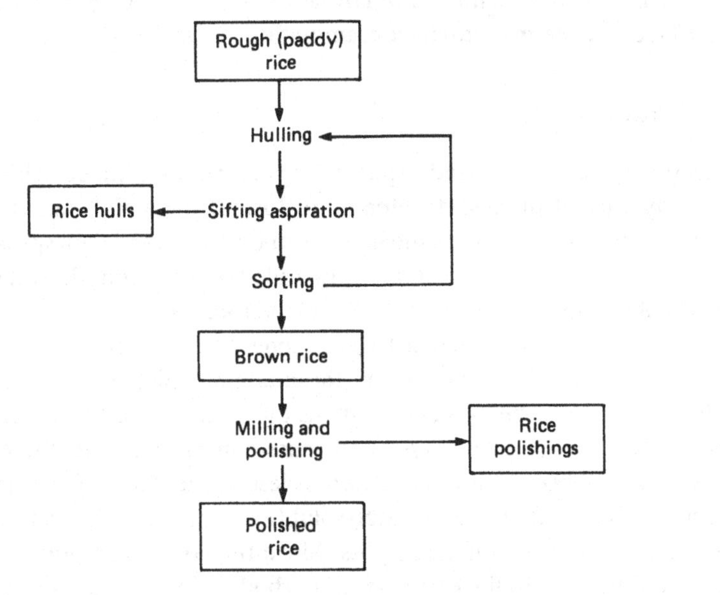

Fig. 13.11. Rice milling.

After the rice has been removed from the field it is threshed and if necessary dried. The hulled paddy rice (rough rice) is now dehulled. It is then subjected to sifting and aspiration which remove the rice hulls. Some unhulled rice still remains and this is separated mechanically from the brown rice. The unhulled rice is returned to the huller. The brown rice is now subjected to a milling process which removes the germ and the outer brown layer. The material is now referred to as unpolished milled rice. This is passed through a brush machine which removes the remaining bran layers and the aleurone layer and removes them as rice polishings. Any broken grains are now removed and the final products is referred to as 'polished rice'.

While much research has been done on the growing of rice, little has gone into the engineering aspects of small rice mills. With multistage milling 70% edible rice is obtained from the paddy of which about 40% is whole-grain polished rice. The rest consists of broken grains of lower value in the market place.

With traditional single-stage milling only 50% edible rice is obtained from the paddy. Half of that is broken grain and the other half a mixture of hulls, polishings and powdered rice. Therefore these mills do not produce a marketable product and the waste even if eaten by the producer's own family is great.

Improper drying, heavy dew and drizzle or high relative humidity in the wet tropics cause fissuring of the grain which also results in low milling yield and incidentally in infestation by field and storage fungi with the risk of spoilage or even mycotoxin contamination.

Parboiling

Occasionally rice is subjected to parboiling before being milled. This is an extremely ancient process developed in Burma, India and Bangladesh. The paddy rice is boiled for some hours in open kettles and then spread on the ground to dry. It is then stored and milled when required. Today, in developed countries, the process is highly automated.

There are several advantages to parboiling: the hulls are loosened and easier to separate. For that reason there is a 3% higher extraction on milling. Moreover, the process toughens the grain and reduces breaking losses. Also, the aleurone layer and scutellum are stuck to the endosperm and on subsequent milling there is less nutrient loss. There is also diffusion of both minerals and water-soluble vitamins into the endosperm which again causes less nutrient loss. Since the colour contained in the bran also diffuses into the endosperm, parboiled rice is easily recognised

Table 13.1 *The effect of milling, polishing and parboiling on rice (per 100 g)*

Component	Brown rice	Polished rice	Parboiled rice
Energy (kJ)	1500	1520	1540
Protein (g)	7.5	6.7	7.4
Fat (g)	1.9	0.4	0.3
Fibre (g)	0.9	0.3	0.2
Ash (g)	1.2	0.5	0.7
Thiamin (mg)	0.34	0.07	0.44
Riboflavin (mg)	0.05	0.03	—
Niacin (mg)	4.7	1.6	3.5

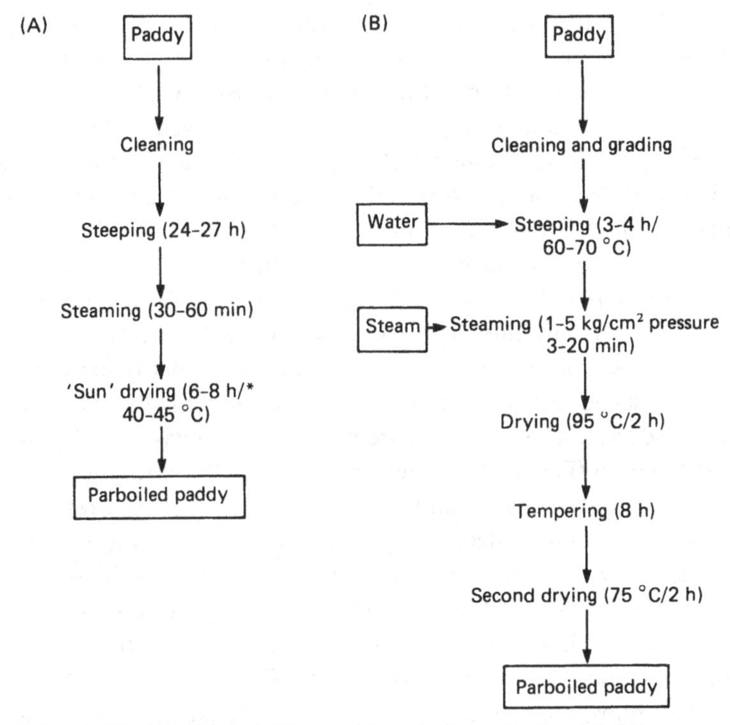

Fig. 13.12. Traditional West African (A) and industrial (B) parboiling of rice.
*Note: at sun-drying temperatures of 40–45°C, rice temperature is about 2–3°C lower due to evaporation. Rice is often dried slowly in the shade to avoid undue endosperm cracking (checking).

by its translucent light-brown appearance. (This must not be confused with the dull reddish colour of brown rice.)

Other advantages of parboiling are that the rice becomes less liable to insect infestation since insects do not have the enzyme system to cope with gelatinised starch. There is also a reduction in mycotoxins if these are present. In canned formulations parboiled rice is preferred because it becomes less mushy on prolonged heating. For the same reason parboiled rice is preferred for institutional feeding because it can be kept hot for many hours. Wheat has also been parboiled and is referred to as 'bulgur'. The effect of milling, polishing and parboiling of rice is shown in Table 13.1. Figure 13.12 shows the flow diagram of the industrial and the traditional West African parboiling process.

Preparation of sugar

Pure cane sugar is prepared in two stages. The first is the manufacture of the raw sugar, the second is the refining process to yield white sugar of high purity. The isolation of the crude sugar usually takes place near the growing area. When the sugar cane arrives at the factory it is first washed to remove any dirt. It is then milled by a combination of knives, crushers and shredders to obtain the maximum amount of juice. The shredded material called 'bagasse' contains about 50% of juice. The juice is filtered off and the bagasse further extracted with water. The bagasse, when dried, is often used as boiler fuel. The juice is now clarified by precipitating impurities with lime. The lime also adjusts the pH and so prevents sugar decomposition during later processing. The juice is now evaporated under reduced pressure in vacuum pans and finally the raw sugar crystallises out. These crystals are removed batchwise or by continuous centrifugation (Fig. 13.13). They are already 98% pure sugar.

The refining of sugar usually takes place near major centres of population, often thousands of kilometres away from the growing area. First the raw sugar crystals are washed to remove the adhering film of molasses. This process is called 'affinition'. The washed crystals are then dissolved and clarified. There are two main methods of clarification. In the first, lime and carbon dioxide are added at 60–80°C and pH 10. Alternatively, lime and phosphoric acid are added to give tricalcium phosphate. This precipitate is difficult to filter and various filter aids are in use. The liquor after clarification is decolorised by the action of bone char (carbon) obtained by anaerobic heating of animal bones. Finally, the refined sugar is crystallised and the mother liquor removed. This process is similar to the crystallisation of raw sugar (Fig. 13.14).

An alternative method for producing sugar in some developing areas (notably India, Asia and East Africa) is the open pan sulphitation process (O.P.S.). The juice is expressed first and mixed with lime. It is then

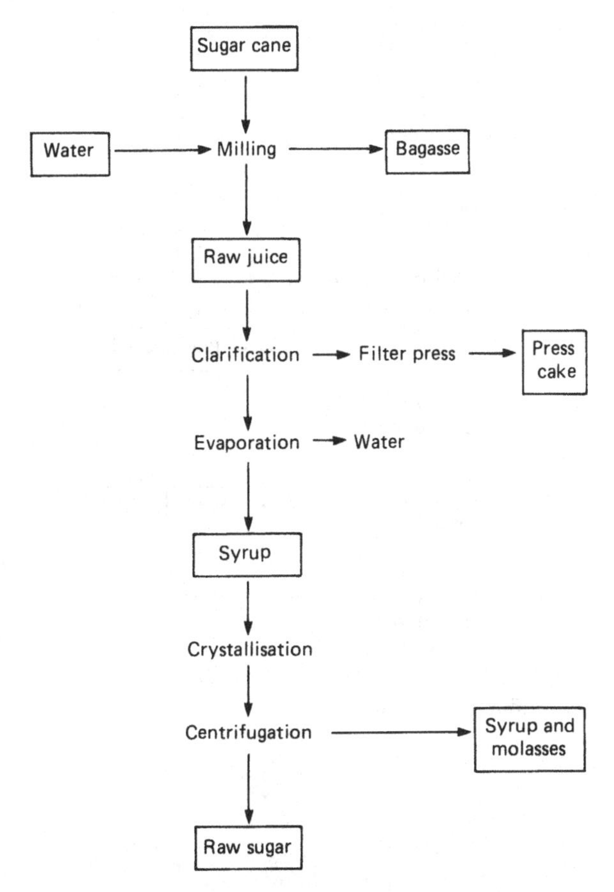

Fig. 13.13. The manufacture of raw cane sugar.

brought to the boil while sulphur dioxide is bubbled through. Two precipitations are normally carried out to remove vegetable debris. The juice is now filtered through cloth and boiled in open pans until supersaturated. It is then cooled and vigorous stirring causes crystallisation. The crystals are then dried.

The advantage of the process is that it requires little complicated and expensive equipment. This can be made locally and operated with little skill.

Disadvantages are that the product is usually brown in colour, tends to be sticky, has a molasses flavour and does not flow easily. The process is also inefficient since only 50–65% of sugar is recovered. The sugar is useful for domestic purposes but not very desirable for industrial manufacture of, for instance, soft drinks, ice cream or sweetened milk.

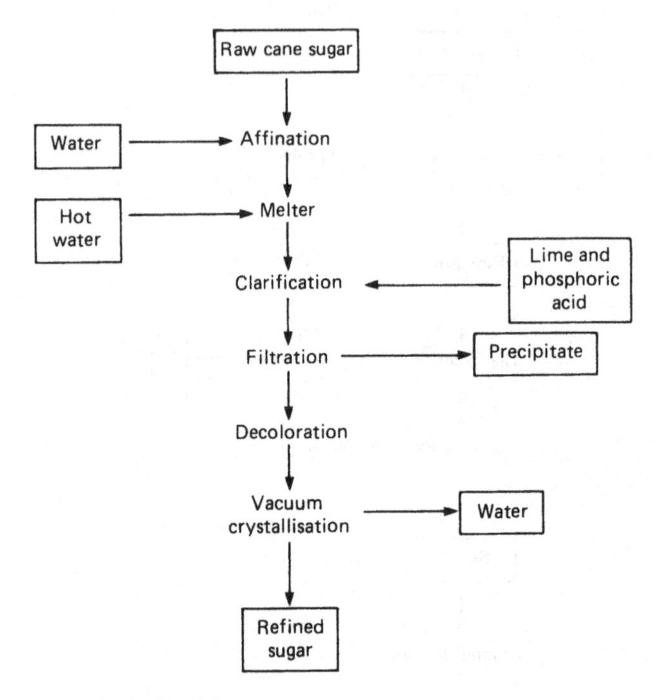

Fig. 13.14. Refining cane sugar.

Fig. 13.15. Crushing sugar cane in India by the traditional method.

Fig. 13.16. Traditional manufacture of jaggery in India.

The traditional manufacture of jaggery is described on p. 166 and illustrated in Figs 13.15 and 13.16.

Protein isolation

Protein is the basic component of every living cell. This means that there are potentially vast amounts of protein available from leaves, seeds, animal products, microbial and many other sources. It also means that man requires considerable quantities of protein and in many parts of the world, particularly the wet tropics, protein foods are scarce.

It is therefore not surprising that a great deal of work has been done to isolate and use protein from many sources both for animal feed and human food.

As an example, about 60% of all soya beans produced are processed industrially and four main products are obtained.

(1) Full fat flour. This consists of the heat-treated and dehusked bean ground into flour.
(2) Defatted soya bean flour. This is the residue left after extraction of fat from the full-fat flour. Since soya beans are a big source of oil a great deal of this material is available in the world market.

Table 13.2 *Analysis of soya beans and some of their products (per 100 g)*

Constituent	Soyabeans	Full fat flour	Defatted flour	Protein concentrate	Protein isolate
Protein (g)	40	41	53	65	92
Fat (g)	22	23	1	0.3	—
Fibre (g)	32	1.7	3	3	0.3
Ash (g)	5	5	6	5	4
PER	—	2.15	2.3	2.3	1.1–1.2
Carbohydrate (g)	7	—	—	—	—

(3) Soya protein concentrate. To produce this the defatted meal is often first heat-treated to make the proteins insoluble. It is then further extracted with alcohol, dilute acid or hot water. This removes sugars and other low molecular weight compounds. In this way the protein content is raised to as much as 70%.

(4) Soya protein isolate. This is the purest type of soya protein available. The defatted meal is dispersed in water in a tank at pH 9 and the coarse material removed by wet sieving. The fine material is then removed by centrifugation. The protein is now precipitated at pH 4.5 to give what is known as 'curd'. This is vacuum-filtered and dried and the protein isolate is obtained as a fine yellow powder. Protein isolates can also be obtained by acid leaching followed by heat precipitation and spray drying (see Chapter 15). An analysis of soya beans and some of the products is given in Table 13.2.

Attempts have also been made to isolate protein from leaves. The tissue of the leaves is first broken down by enzymes, rapid freezing, wet milling using hammer, roller or pinmills or by ultrasonic vibration. The material is then filtered and the fibrous residue (usually more than half of the solids) is removed. This is suitable for microbial fermentation or fodder for ruminants. The juice is precipitated either by acid or heat and the precipitate removed by filtration. The filtrate contains significant amounts of sugar, water-soluble salts and amino acids and can also be used for microbial fermentation. The residue is washed and either canned as a wet cake or dried.

This process, or one like it, can be carried out on a large industrial scale, at village level using electric or diesel-driven mills or in the laboratory with a meat mincer or blender. Fifty per cent of the leaf protein is fairly easily extractable in this way and isolate containing 10–30% protein on a dry weight basis has been obtained.

The fact that the isolation of leaf protein for food has so far not given rise to a large industry in spite of the huge amounts available suggests that there are problems. The protein content of fresh leaves is very low (1–4%). Although it has been quoted as between 15 and 30%, this is on a dry weight basis and economic considerations must apply to an extraction of leaves of natural moisture content and not of dry ones. Dry leaves are difficult to extract. The leaves should also not contain mucilage which would make them difficult to process. There should be no toxic elements or undesirable flavours and the leaves should be available in bulk for most of the year. For this reason industrial processing waste is useful but fresh market waste is not. Care must also be taken that any excessive amounts of fat in the leaves do not produce rancidity. Finally, the usual leaf protein isolate has a bright green colour and smells very much like new-mown hay.

Basically, for the process to be economical, sufficient amounts of suitable fresh green foliage must be available at the processing plant for most of the year. Since tree leaves tend to have a higher protein content than those of herbs, trees can be grown specially and the leaves cut periodically. Alternatively, protein isolation can be associated with some other operation such as the processing of cotton, sugar cane, banana or jute. As with all methods of protein isolation, energy costs, i.e. those of firewood, charcoal, oil, gas or electricity are extremely important.

Isolation of protein from animal products has the advantage that there is no fibre to be removed but animal products are often more expensive. An important source of animal protein is fish meal. Approximately 40% of the world's catch is converted into it. The fish is dried and ground and used for poultry and pig feeds. The largest exporters of fish meal are Peru, Norway and South Africa. Fish meals usually contain 10% water, not more than 6% fat and 4% salt. Protein content is usually high, of the order of 65%.

There are three main types of fish protein concentrate (F.P.C.)

- Type A – almost colourless and tasteless powder with a maximum total fat content of 0.75%.
- Type B – powder without specific odour or flavour limits. It has a definitely fishy flavour, the maximum fat content being 3%.
- Type C – normal fish meal.

To manufacture F.P.C., fresh fish is first rinsed and then ground. The

Table 13.3 *Composition of some milk products (per 100 g)*

Product	Water (g)	Energy (kJ)	Protein (g)	Fat (g)	Total carbo-hydrate (g)
Full cream milk	88	272	3.5	3.5	4.7
Skim milk	91	142	3.4	0.1	5.0
Full cream sweetened condensed milk	26	1360	8	9	56
Skim sweetened condensed milk	27	1140	10	0.3	60
Full cream dried	3	2050	26	26	40
Skim dried	4	1510	36	1.3	53

mass is agitated with isopropanol and some of the liquid is removed at room temperature in the first extractor. The material is centrifuged and then heated in a second extractor at 75°C under agitation. The meal is now almost dehydrated and the fat content reduced from 5% to almost 1%. The liquid and the solvent are recirculated. The resulting cake is vacuum-dried, ground, sieved and packed. Types A and B resemble normal fish meal C and a low fat content is desirable to avoid rancidity.

Protein content of type A is as high as 80% and of types B and C of the order of 65%. The amino acid composition is similar to that of fish. At present, the material is produced only in small quantities for aid programmes, but is not sold commercially. It can be incorporated into bread, biscuits, soups and stews. Types B and C are best in foods where fish flavours are not objected to.

The development of milk concentrate sprang originally from the desire to preserve the food value of the milk while reducing its bulk. Today, dried milk powders are important protein sources particularly for infants. The raw material for all these products is normally cow's milk (full cream milk). When the fat is removed, usually for butter manufacture, the remaining product is skim milk. Both of these types of milk can be further processed by removal of water to give condensed milk (whole or skim). Sugar may be added to give a total solids content of 75%, 55% being sucrose. A further removal of water may take place either using steam-heated rollers (roller powder) or by spraying concentrated milk through hot air in a spray drier (spray powder) (see Chapter 15). Table 13.3 gives the analysis of some milk products.

Apart from the protein sources mentioned above residues after extraction of oil from groundnuts, oilseed, cottonseed, coconut, sunflower and rape seed have been used. Perhaps those from oilseeds are the most promising of all for cheap protein foods for developing countries. Single cell (see Chapter 14) leaf and fish protein concentrates are at present not cheap and often not stable enough for them to be used extensively.

Production of palm oil and palm-kernel oil

The oil palm fruit yields two types of oil. The outer fibrous layer provides palm oil, the inner hard nut contains palm-kernel oil (Fig. 7.11). These oils are obtained as follows.

Palm oil

The fruit is boiled in water to soften it. It is then pounded and again heated in excess water. The pulp is squeezed and the oil skimmed off. The oil is heated again to expel any occluded water (Fig. 13.17).

It is also possible to soften the pulp by fermentation assisted by pounding. Again, the oil is finally separated from hot water.

A more advanced method consists in first steaming or boiling the whole bunches of the fruit to destroy the enzymes. The fruits are then separated

Fig. 13.17. Extraction of palm oil as used by small-scale processors in Nigeria. As the water boils the oil rises to the top. When cold it is skimmed off.

Fig. 13.18. Palm oil extraction in West Africa using a simple hand press. The oil is collected in a kerosene tin. Capacity is about 1 ton of oil per week. (Courtesy UAC International.)

and digested at about 110°C. The oil is then obtained by expression (Fig. 13.18) or on a large industrial scale by centrifugation. If facilities are available the oil is then neutralised, deodorised and de-watered.

Palm-kernel oil

Here the hard kernels are cracked open with a heavy stone or using a centrifugal or hammer mill (if available). They can then be extracted. Industrially, the kernels are ground using flaking rollers and finally extracted with a hydraulic or screw press or by repeated solvent extraction. Using the traditional methods only 40–50% of the palm oil present is extracted. The rest is thrown away with the fibrous residue. Also, the oil obtained is of inferior quality and would not be acceptable in the world market. This is mainly due to the free fatty acid content which is far too high. This results in off-flavours.

The oils are unusual amongst plant oils in being high in saturated fatty acids. Most seed oils consist mainly of unsaturated fats. Palm oil contains about 50% saturated (mainly palmitic acid) and 50% unsaturated fatty acids (mainly linoleic and oleic acids).

Fig. 13.19. In the villages on the Indian subcontinent butter is still churned by the traditional method.

Manufacture of butter and margarine

Butter contains 80% butterfat, 0.4% protein, 1% non-fat milk solids, 16% water and up to 3% added salt. The colour of butter is variable and depends on carotene content. It may be adjusted artificially by the addition of annatto. The physical properties which affect the spreadability of the butter are largely governed by the proportion of solid fat to liquid oil. This depends on both the temperature and the fatty acid composition which is subject to seasonal variations.The traditional manufacture of butter is by batch process. The cream is either skimmed off from the top of the milk or separated by means of a centrifugal separator. The cream is then briefly ripened. This stage is equivalent to the rather more prolonged ripening in cheese fermentation. The ripened cream is then churned and in the resulting phase inversion (from an oil in water to a water in oil emulsion) butter and buttermilk are produced (Fig. 13.19). The butter is now washed and is ready for consumption.

On the Indian subcontinent butter is often heated and skimmed. The product is referred to as ghee. In the process water and protein content are reduced to a trace and the energy value increases from 3000 kJ/100 g to 3700 kJ/100 g. The keeping properties are improved.

Modern continuous methods of butter manufacture are basically of two types. In the first, cream of 35–40% fat content is passed as a very thin layer through a cylinder. Violent agitation inverts the emulsion within a few minutes as compared to over an hour in the traditional churning process. With the second method, the cream is first pasteurised and then evaporated to a fat content of approximately 80%. It is then homogenised. In a second separator, 98% milk fat is obtained which is passed to a blending tank where water, curd and colour are added. The resulting butter is now worked and finally chilled.

It may be necessary to adjust the pH of butter and this is done with sodium carbonate. In order to remove bacteria, the butter may also be heated to 88–90°C in a plate heat exchanger (see Chapter 15). It is then cooled to 67°C and held to allow for crystallisation. This will give the right proportions of solid fat crystals to liquid oil on which the spreadability of the butter will depend.

For many years margarine has been sold as a cheap substitute for butter but the last 20 years have seen great progress in its composition and manufacture.

Basically, margarine consists of 80% fats or oils. Animal fats used include beef, pig or mutton fat, whale oil and fish oils. If there is religious objection to the use of animal fats these are omitted. The plant fats are obtained mainly from soya, cottonseed, coconut and oil palm. Margarine also contains 16% water, sodium chloride and between 3 and 10% milk or milk products. Vitamins A, D and E are also often added. Other additives may include colour, such as carotene or annatto, flavours (e.g. diacetyl), emulsifiers, such as lecithin, soya or wheat flour, preservatives (boric, benzoic or sorbic acids if permitted) and antioxidants.

In manufacture the crude oil is bleached and deodorised and a proportion may be hydrogenated to give a suitable melting point in the final blend. The oils are then blended in correct proportions and passed to the premix tank. Here emulsifier, salt solution, vitamins and pasteurised skim milk are added. The latter has previously been inoculated with starter cultures of *Streptococcus lactis* or *S. cremonis* and matured for about 20 h at 22°C. From the premix tank the blend is passed to a scraped-surface heat exchanger (a votator), a machine where emulsification and conditioning are carried out in a continuous operation. Suitable adjustment of speed and cooling allows the manufacture of margarines of different textures (Fig. 13.20).

Low-energy margines have also been produced. While butter contains, per 100 g, 2990 kJ and margarine 3010 kJ, low-energy margarine contains

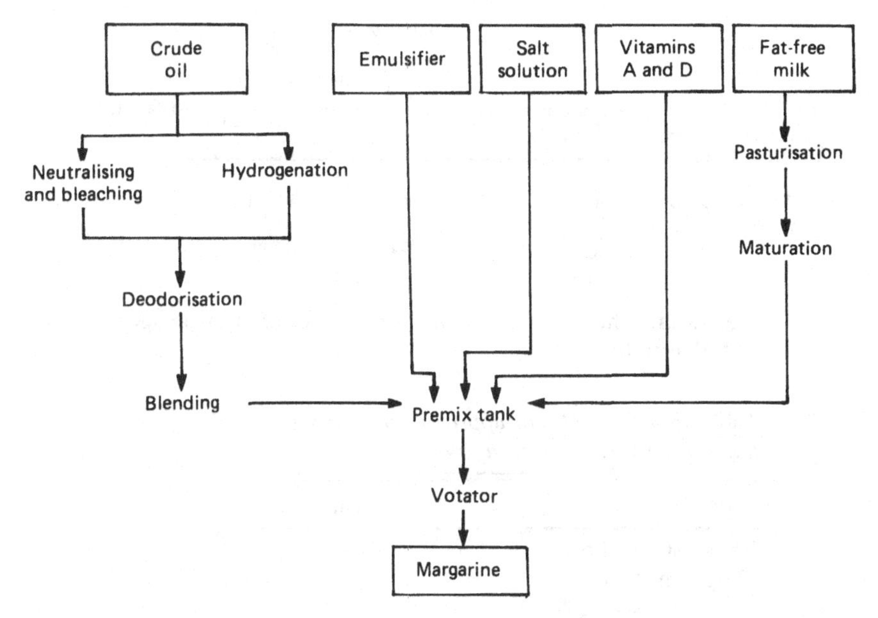

Fig. 13.20. The manufacture of margarine.

only 1550 kJ. This low-energy content has been achieved by increasing the water content of the emulsion.

The nutritional properties of butter and a well compounded margarine are very similar. Because margarine is made from a mixture of fats hardened to varying degrees it need not have a distinct melting point. A good margarine should have a wide melting range allowing it to be spread easily over a wide range of temperature.

Heat treatment

The object of cooking is mainly to improve the appearance, flavour and digestibility of the food. The effect of heat on nutrients has been considered in Chapter 2. Foods are very complex materials and the effect of heat depends both on the food and the condition of heating.

All heating processes are based on energy input and all waves are electromagnetic disturbances of the same type. The visible spectrum is perceived by the eye, heat is received by nerve endings in the skin but man has no receptors for electric waves. Figure 13.21 shows the wavelengths of the various types of electromagnetic disturbance.

The heat produced in a standard oven at 350–400°C has a wavelength of about 5×10^{-5} m. Infrared heating is similar to conventional heating. It heats the product from the outside but does not heat the surrounding air.

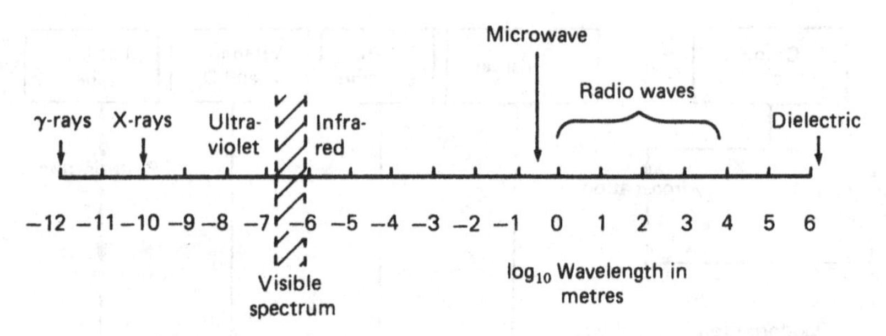

Fig. 13.21. The wavelengths of various types of electromagnetic disturbance (m).

Table 13.4 *Examples of approximate cooking times in a microwave oven*

Food	Time
300 g smoked fish	45 s
225 g chicken	45 s
170 g steak	45 s
350 g white fish	60 s
soup	1 min
350 g red meat	10–15 min
5 kg goose, turkey	35 min

Heat penetration depends on wavelengths. With high frequency (microwave) heating an energy change takes place within the whole volume of the product. Because heat loss occurs at the surface, the centre of the food eventually reaches a higher temperature than the surface and there is no crust.

The cavity magnetron which produces the microwaves was invented in 1940 in the UK and became radar's most closely guarded secret. In 1945 an American scientist testing radar equipment found that a chocolate bar in the ray's path had melted. He sent for a bag of popcorn kernels, placed them before the aerial and saw them turn into fluffy popcorn.

The advantages of microwave heating, which is becoming more popular both in industry and in the home, are low running costs and very rapid cooking time, resulting in less nutrient loss (Table 13.4). Disadvantages are that it is not possible to cook a complete meal at the same time unless it is already pre-cooked and subsequently frozen. Because foods vary in composition, particularly water content, they absorb microwaves at different rates and have to be cooked separately. Eggs cannot be boiled as the steam produced bursts the shell. Deep frying using microwaves may result in overheating of the fat causing it to burn. Roasts and bread look very pale because there is no crust formation and microwave heating must be used together with conventional or infrared processes. Because microwave heating interferes with radio transmission a very restricted number of wavelengths only are allowed (two in the UK).

Cooking of meat and fish

Meat contains more protein than plant products and therefore shows effects typical of protein changes. The haeme pigments, the red colouring matter, turn from red to grey and the proteins coagulate. There is no need to reach the boiling point of water because these changes already take place on simmering.

Many flavouring substances contained in the meat are easily leached and therefore a minimum of water should be used. As the proteins become denatured, water is lost. Examples of water loss from meat are muscle 10%, brain 1%, heart 20%, tongue 5%, kidney 20% and liver 25%. There may also be a fat loss from the tissue and as a result the protein content increases.

Frying can take place in an oil bath, by contact frying (i.e. single sided in a pan or double sided between two heated plates) or in an oven.

On frying meat the heat reaches the meat directly rather than through the water. The ultimate surface temperature of the meat is higher than on boiling and Maillard reactions occur. Water loss is slightly higher than on boiling and excessive dessication is prevented by basting with fat.

There is also a reaction with the heating medium. In air oxidation takes place and in an oil bath the oil is taken up. The temperature at the centre lags behind that at the surface and only reaches a maximum when cooling has already begun (Fig. 13.22).

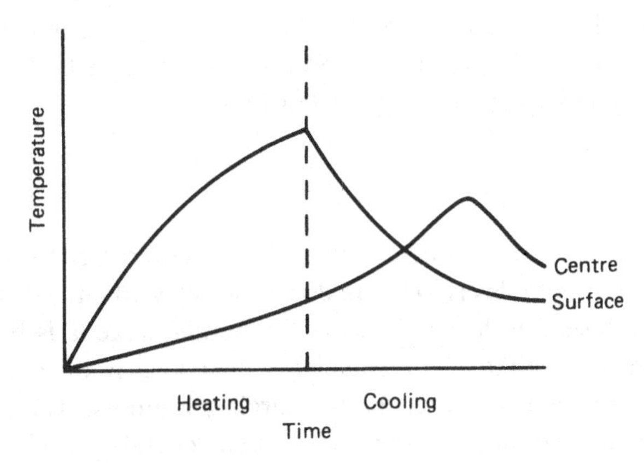

Fig. 13.22. Heating and cooling during frying and baking.

Table 13.5 *Water content of some raw and cooked vegetable foods (g per 100 g)*

Food	Raw	Cooked
Peas	14	70
Potato	75	78
Cabbage	87	85–90
Arrowroot	12	90
Banana	70	90
Soya bean	10	70
Rice	14	70

Table 13.6 *Stability of vitamins on heat processing*

	Vegetable			Meat		Egg		Fish	
Vitamin	Root	Leaf	Seed	Boiled	Fried	Boiled	Fried	Boiled	Fried
A	S	S	S	S	S	S	S	—	—
D									
E	S	S	S	FS	FS	S	S	S	S
K	—	—	—	—	—	S	S	—	—
B_1	FS	VU	U	VU	FS	S	FS	S	FS
B_2	U	VU	U	U	FS	S	FS	S	FS
B_5	U	U	U	VU	FS	S	FS	FS	FS
B_6	VU	VU	VU	VU	FS	S	FS	S	FS
Niacin	U	VU	U	VU	FS	S	S	S	FS
B_{12}	—	—	—	FS	FS	S	S	S	S
M	VU	U	VU	—	—	S	U	S	S
C	VU	VU	VU	FS	FS	—	—	—	—
H	U	U	U	FS	FS	—	—	FS	FS

Stable (S), 0–9% loss; fairly stable (FS), 10–24% loss; unstable (U), 25–39% loss; very unstable (VU), 40–70% loss.

Fish has less flavour than meat to start with and the flavouring substances are more easily leached. Therefore fish must be boiled very carefully and is normally steamed, grilled or fried.

Cooking vegetables

Cellulose and raw starch are almost indigestible and therefore the prime object of cooking vegetables is to break down the cell wall and gelatinise the starch. Cellulose is little affected but the middle lamella is broken down by the heat. Unlike meat and fish which lose it on cooking, vegetables tend to take up water as the starch gelatinises. Table 13.5 shows the water content of some raw and cooked vegetable foods.

During cooking chlorophyll pigments may turn from green to yellow or brown. The carotenoid pigments are little affected but anthocyanins are relatively unstable. Banana and apples may turn pink on heating due to a change in anthocyanin.

While minerals remain unchanged on heating (they may of course be leached) the effect on vitamins is very variable. This is shown in Table 13.6.

Baking and the manufacture of snack foods

Perhaps more than any other branch of food technology, bread baking in the tropics shows enormous differences in method and equipment. Therefore it is best to describe first the process that was usual in the 1950s and 1960s and then consider some of its variations (Fig 13.23).

Flour was delivered from the mill in 2-ton bulk lorries and conveyed to the silos. When required it was mixed with yeast, salt and water in a planetary mixer (Fig. 13.24). The resulting dough was fermented at 26°C for about 3 h, divided into suitable pieces, rounded on an umbrella moulder and proved. The prover is a temperature- and humidity-controlled cabinet in which the round dough balls travel on shelves fitted with fabric pockets, one for each dough ball. After the proof the dough pieces were moulded into cylinders and placed into tins. These were passed to the final prover where the dough rose a little, perhaps to the edge of the

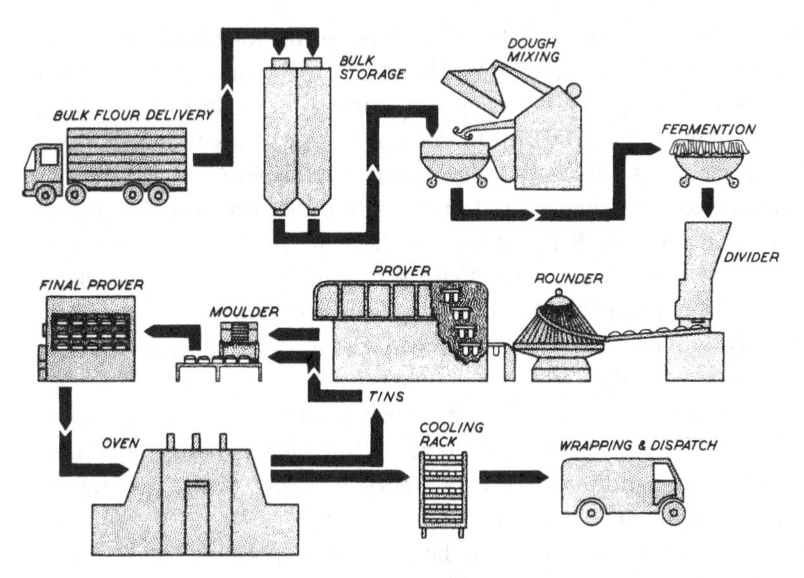

Fig. 13.23. The typical bread-making process (see text).

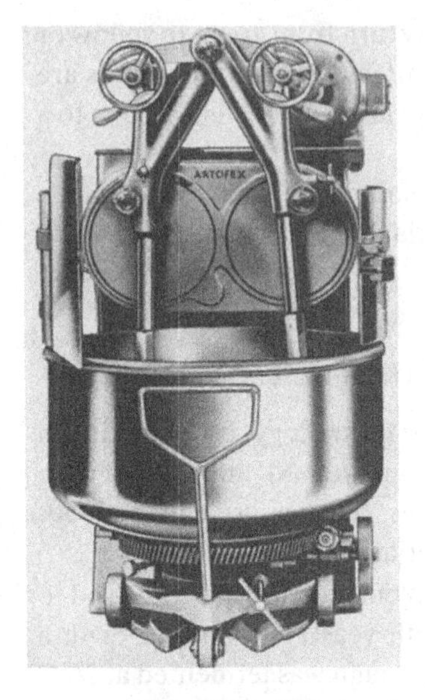

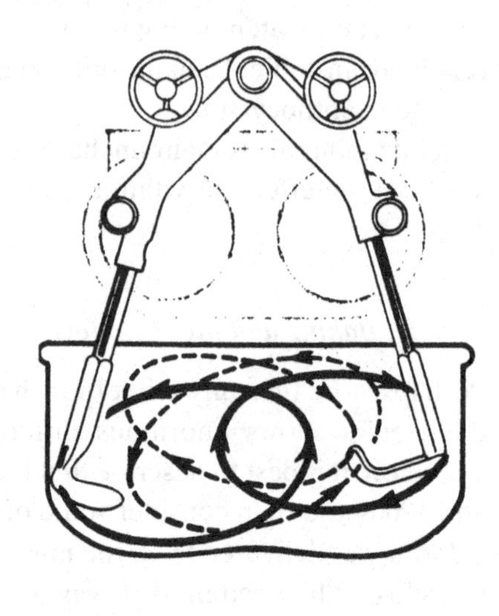

Fig. 13.24. The Artofex is a versatile planetary dough mixer which imitates the baker's arms. As the action of the two arms keeps the mixing to the centre of the bowl, small as well as large batches can be dealt with efficiently. (Courtesy Artofex Engineering Works Ltd.)

tin. The dough was then baked and the bread de-tinned and cooled on racks while the tins were recirculated.

What then are the modifications? Instead of the flour arriving in bulk lorries it may arrive in sacks, enamel basins or calabashes. Mixing is often done in bowls, troughs or on table tops. High-speed mixers may be used (Fig. 13.25) by means of which the dough receives a great deal of work input. In this way fermentation time can be greatly reduced.

Fermentation is often omitted in the case of flat bread. Examples of these are chapati, tortillas and kissra. These are flat breads raised only slightly by steam. The chapati (Figs 13.26, 13.27) is made from wheat flour and eaten in the Far East. The tortilla is made from maize and is the traditional food of Central America, while kissra is made from sorghum and eaten in the Sudan.

The divider and rounder are often dispensed with, the dough being divided and shaped by the baker's hand. In West Africa doughs are often worked by a doughbrake, a set of rollers through which the dough is passed repeatedly. This gives better loaf volume.

Provers are expensive and often the dough is simply allowed to rise in a warm place.

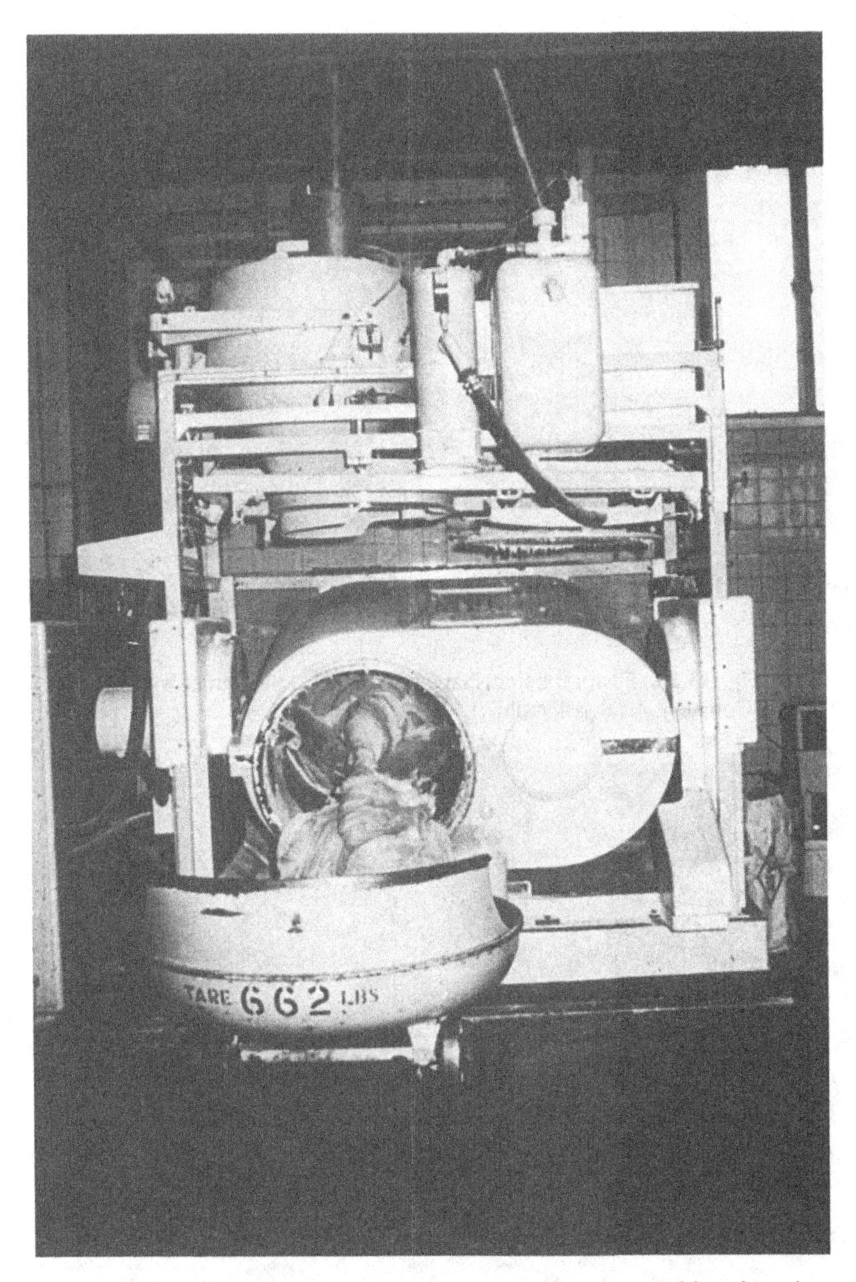

Fig. 13.25. A high-speed batch mixer in the open position dropping the dough into the bowl. (Courtesy Tweedy of Burnley Ltd.)

Fig. 13.26. Chapati being shaped and baked on a small wood stove. (Courtesy A. B. Chaudhri.)

Fig. 13.27. Flat bread can now be made with automatic machinery. The round dough pieces (right) are flattened and move straight into the oven (left). (Courtesy Baker Perkins Holdings plc.)

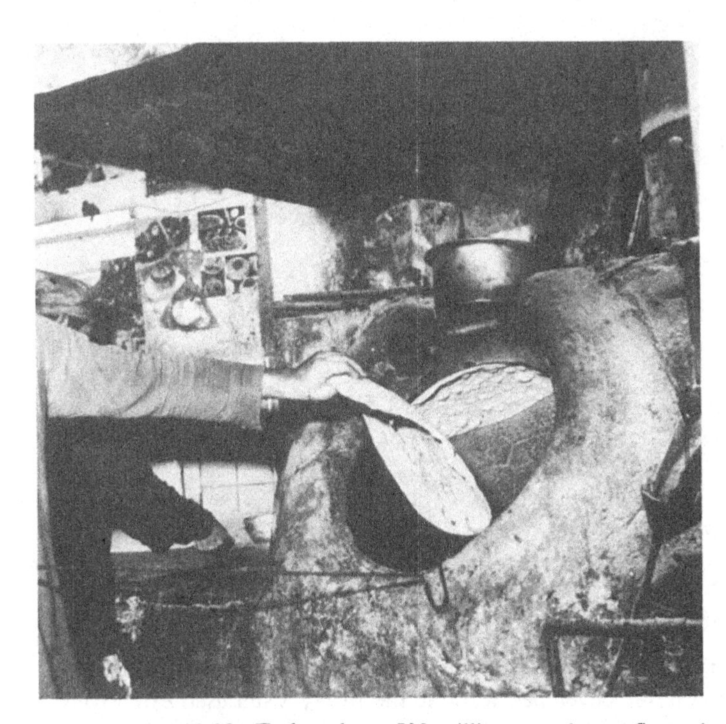

Fig. 13.28. Today about 500 million people eat flat unleavened bread. Here thin Iranian bread called 'taftoon' is taken from the circular clay oven. (Courtesy H. Moaven Shahidi.)

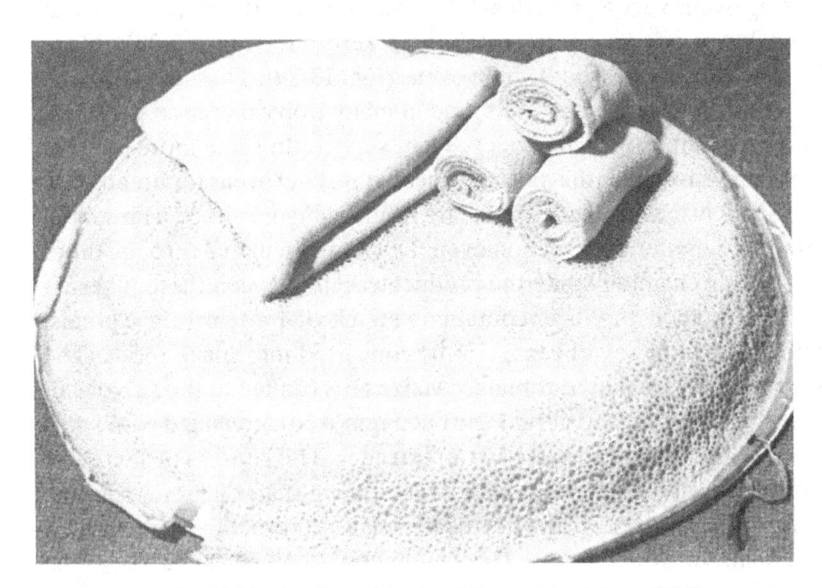

Fig. 13.28a. The Ethiopian bread 'enjara' is prepared from T'ef (see p. 109). (Courtesy A. Habtamariam.)

Fig. 13.29. This type of chain-travelling oven for baking biscuits, popular in the developed world in the 1930s, is still used extensively in some tropical countries.

Baking ovens vary a great deal. The commonest is the clay oven which is fired by wood, dried dung cakes or kerosene (Fig. 13.28). More advanced are the brick-built chain ovens (Fig. 13.29). These are obsolete in developed countries but still found in many tropical areas e.g. South-East Asia where they still give excellent service. In some more affluent countries of the tropics the most advanced travelling ovens for bread (Fig. 13.30) and biscuits (Fig. 13.31) are found. Figure 13.32 shows the diagram of a modern convection oven. Hot air is circulated through tubes in the baking chamber while the product travels between these tubes.

Many raw food products containing starch or protein (e.g. cereals, potato, cassava or soya beans) can be converted into snack foods. The most common cereal used is maize. Maize grits (milled to the size of salt grains after the bran and germ have been removed) are mixed with some salt and a little water and passed to an extruder. This consists of a screw of constant overall diameter but with an increasing shaft diameter rotating in a close-fitting, grooved and heated barrel. The end of the barrel is fitted with a plate which is perforated by a number of small holes (the die). The increasing pressure and temperature during extrusion disrupts the starch as it gelatinises and denatures the protein. The material is extruded at

Fig. 13.30. This travelling oven bakes 2400 large loaves every hour. (Courtesy Baker Perkins Holdings plc.)

Fig. 13.31. A modern band oven showing biscuits leaving the baking chamber. (Courtesy Baker Perkins Holdings plc.)

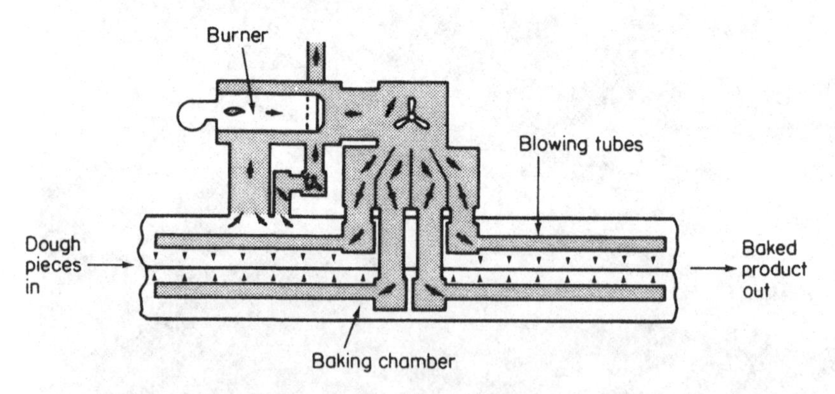

Fig. 13.32. Diagram of a modern convection oven. (Courtesy Baker Perkins Holdings plc.)

about 150 bar (2250 p.s.i.) and 150°C through the die. As the mix emerges into the atmosphere the drop in pressure causes the water to evaporate, the resulting steam expanding the product. Rotating knives at the die cut it into suitable pieces and the cooling to room temperature makes them set. The product is further dried to about 1% moisture content at which it

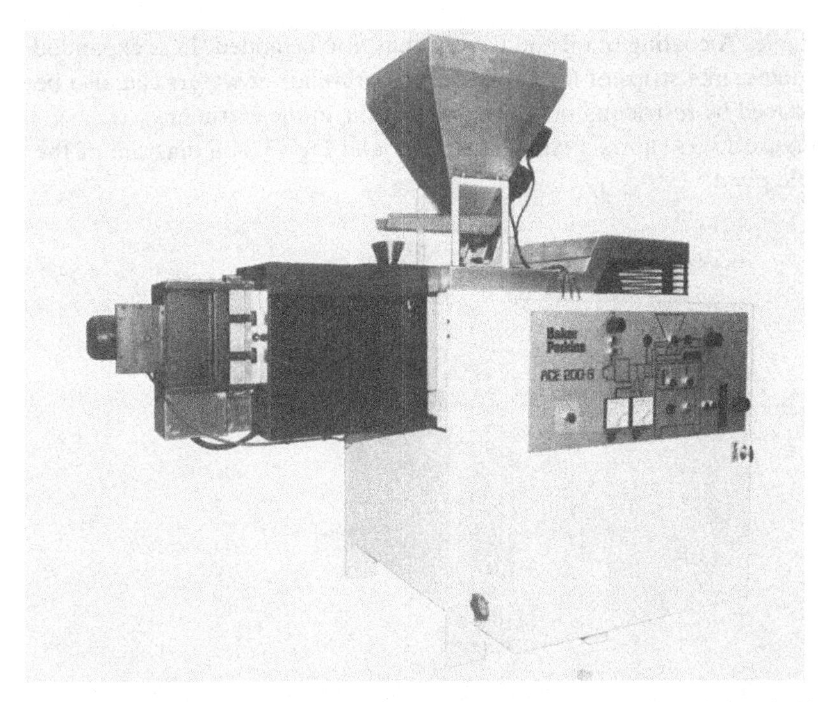

Fig. 13.33. A cooker-extruder plant for the production of expanded snack foods. The hopper at the top feeds via the slide into the small funnel on the top of the barrel. The housing on the left contains the screw, the die and the nozzle with its two-bladed cutting knife at the far left. (Courtesy Baker Perkins Holdings plc.)

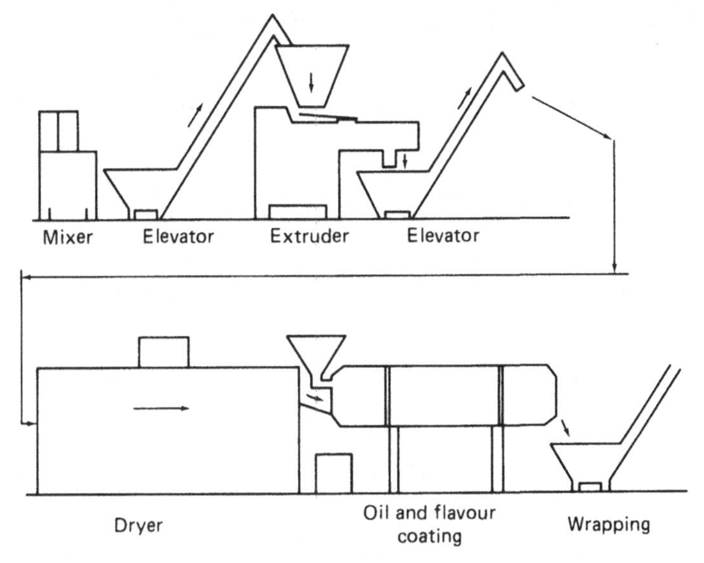

Fig. 13.34. Typical snack food extrusion plant.

is stable. A coating of oil and flavour may also be added. Less expanded products, like strips of flat bread (extruderbread) or wafers can also be produced by restricting pressure and heating in the extruder.

Figure 13.33 shows a typical extruder and Fig 13.34 a diagram of the whole plant.

14

Food fermentations

In a tropical environment methods of food preservation common in developed countries are often not freely available or even impossible. Since microorganisms grow well in the warm climate it is not surprising that traditional food fermentations are found very frequently.

There are basically three types of fermentation. The most primitive, the so-called 'wild fermentation', depends on microorganisms which accidentally fall into the batch. Depending on air- or seed-borne microorganisms is always uncertain since occasionally the wrong organisms will start to work and the mix will be spoilt. Examples of wild fermentations are the traditional preparation of tempeh, kenkey or pito beer.

With more advanced systems 'starters' are added. These are usually the remnants of a previous successful batch which are added at a level of 1 or 2% to the new one. In one instance, the author saw a brewer making pito beer in Nigeria transfer, with great ceremony, a coil of rope impregnated with microorganisms from the previous brew to a new one. He pretended it was magic, but this coil was no more than a successful starter culture.

The most advanced type of fermentation involves the addition of a pure culture. The following are some examples of fermentation to illustrate the range.

Soya bean fermentations

Tempeh is an Indonesian food made entirely of soya beans, and used as a main dish. The soya beans are cooked, dehulled and inoculated with one of several species of *Rhizopus*. The beans are placed on trays and subjected to aerobic fermentation for 1–3 days at 30°C. The product is referred to as raw tempeh cake. For use, it is sliced and fried or baked.

Fig. 14.1. In Tempeh production the inocculated soya beans are wrapped in cloth and allowed to ferment on wooden trays (on trestle table centre). The raw tempeh cake ready for sale is shown in the vessel (front, right). (Courtesy R. E. Muller.)

During fermentation, both fat and thiamine decrease, the latter by about one-third and riboflavin and niacin increase by a factor of three. Fibre content increases by about 3.7–3.9%. Tempeh has also been made from wheat and cassava where protein content is increased by a factor of 6–7. It is a very rich source of vitamin B_{12} which is normally present only in foods of animal origin (Fig. 14.1).

Miso paste has a meat-like taste and the consistency of peanut butter. It is most important in Japan where it is used in the preparation of vegetables and soups. There are many different types of miso which vary in composition and time and temperature of fermentation. With one particular method, washed rice is steamed and then cooled to 35°C. It is then inoculated with *Aspergillus oryzae* and incubated for 50 h at 27°C. The resultant product is referred to as koji. Soya beans are now crushed and soaked for 2–5 h. They are steamed under pressure and cooled. Mixed with koji and salt they are inoculated with *Saccharomyces rouxii* and fermented for two months at 35°C. The product, miso, is then ripened for one week at room temperature, mashed and blended. Protein content ranges from 11–21% and during fermentation vitamins B_2 and B_{12} are produced.

Soy sauce is a brown liquid with a salty meat-like flavour. It is produced

Table 14.1 *Composition of soya beans and some of their fermented products (per 100 g)*

	Water (g)	Energy (kJ)	Protein (g)	Fat (g)	Total carbo hydrate (g)	Crude fibre (g)	Ash (g)
Seeds, raw	10	1685	35	18	35	5	5
Seeds, cooked	70	543	10	6	10	1.5	1.5
Tempeh	60	698	15	8	10	3	1.5
Miso	50	715	10	4	20	2	10
Soy sauce	63	284	6	1.5	10	0	20

from soya beans often with the addition of wheat. It is a popular flavouring in China and Japan. The soya beans are first soaked in water and cooked. If wheat is used it is first roasted to give a slightly charred flavour and then milled. *Aspergillus oryzae* is now grown on the bean–cereal mixture at 25–35°C for about two days. This is now stirred into 1.5 volumes of salt solution and allowed to ferment for as long as three years although fermentation can be as short as one month. The ferment is now filtered by hydraulic pressure, briefly heated to 70–80°C, pasteurised (see Chapter 15) and bottled. The composition of soya beans and some of their products is given in Table 14.1.

Acid fermentations

Idli is a fermented, steamed bread-like cake and an important source of protein and energy in the diet of many southern Indians. Because idli is easily digested it is often consumed by infants and invalids. It may be consumed twice a day for breakfast and supper, the average person consuming two or three cakes at a meal. The cakes are steamed and are about the size of a poached egg. Indeed they can be prepared in a standard egg poacher. Amongst the fermentative organisms are *Leuconostoc mesenteroides*, *Lactobacillus fermentum* and *Streptococcus* spp. The preparation of idli is given in Fig. 14.2.

Ogi is a sour porridge made from maize or sorghum (Fig. 14.3). It is very popular in West Africa and the traditional weaning food in Nigeria. The predominant acids formed during fermentation are lactic, acetic and butyric. Ten volatile acids have also been identified. There is loss of many nutrients in ogi preparation mainly of protein, calcium, iron and some B-group vitamins, although riboflavin and energy are increased. Soy-ogi has

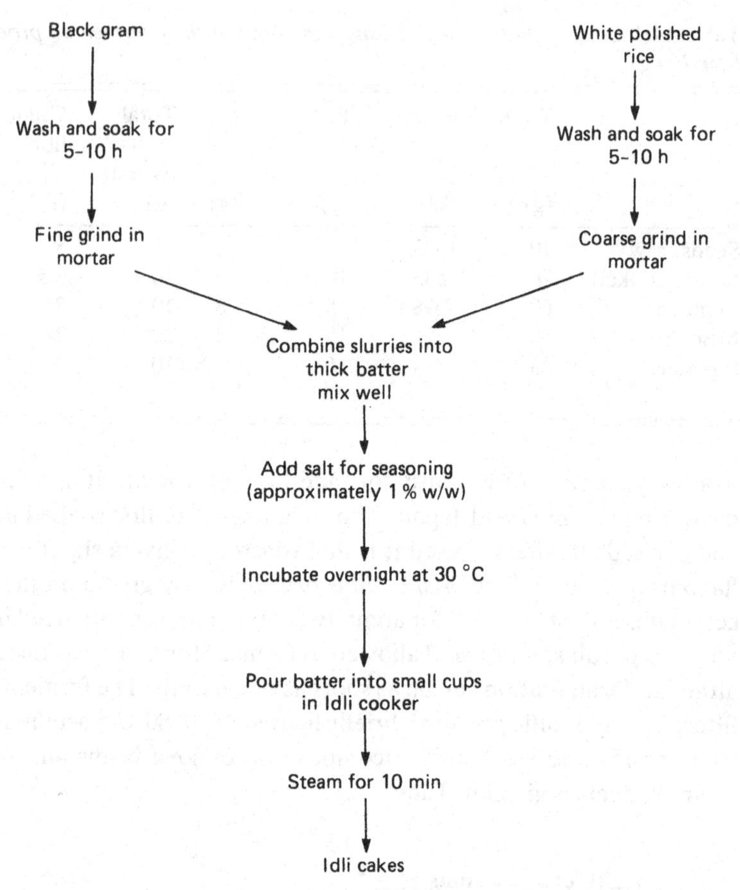

Fig. 14.2. Flow diagram for the preparation of idli.

been prepared industrially by the addition of 10% soya flour. Fermentative organisms include *Lactobacillus plantarum*, *Streptococcus lactis* and *Saccharomyces rouxii*. The flow diagram of the traditional preparation of ogi is given in Fig 14.4.

Kenkey is a fermented maize dumpling, the staple food of the Ghanaian. Steeped maize is ground and made into a dough called aflata. This is fermented for 2–3 days and divided into two equal portions. One remains uncooked, the other is heated (Fig 14.5). Both are then mixed together, formed into dumplings, wrapped in leaves and boiled. They are then cooled and ready for consumption (Figs. 14.6, 14.7). Cooking the aflata is very strenuous work and a satisfactory product can be produced using pregelatinised starch. The mechanisation of kenkey manufacture has also been studied. While wrapping the maize dumplings into

Fig. 14.3. Wet sieving of sorghum in the preparation of ogi (Nigeria). Saddle stone on the left. (Courtesy E.O.I. Banigo.)

polythene or aluminium foil was more hygienic, the flavour imparted by the various leaves used traditionally for wrapping was important. The flow diagram of the manufacture of kenkey is given in Fig. 14.8. Fermentative organisms include *Aspergillus* spp., *Penicillium* spp. and lactic bacteria.

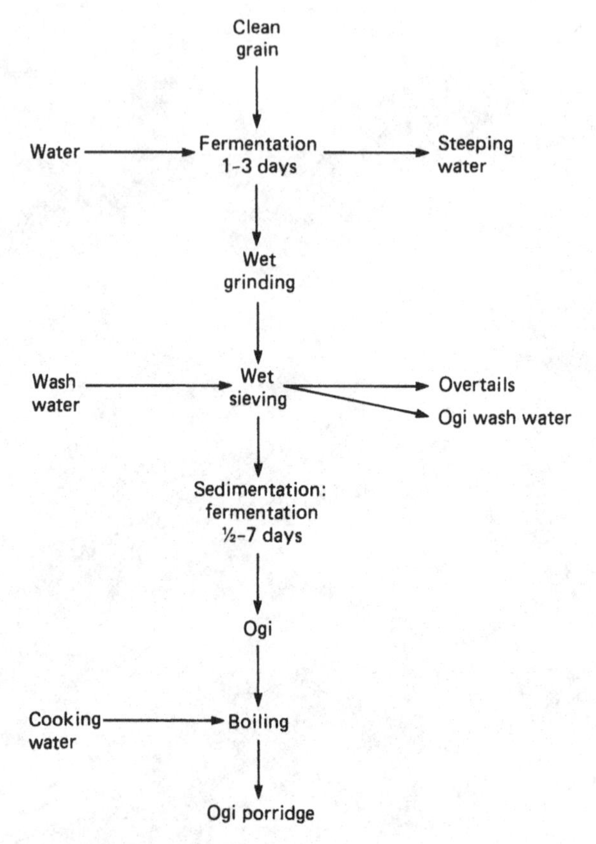

Fig. 14.4. Flow diagram of the traditional preparation of ogi.

Alcoholic fermentation

Of all the various fermentations, alcoholic fermentation was the first recorded and studied. Osiris among the Egyptians, Bacchus among the Greeks and Noah according to the Israelite tradition, taught men the art of cultivating the vine and making wine.

During fermentation glucose is converted by yeast into ethanol and carbon dioxide.

$$C_6H_{12}O_6 \rightarrow 2\,C_2H_5OH + 2\,CO_2$$

Alcoholic beverages may be divided into beers, wines and spirits. The first are made by fermenting cereals and the second by fermenting any plant juice which contains fermentable sugars, i.e. the juice from grape, banana, mango, pineapple, rafia palm, oil palm or the juices from the

Fig. 14.5. Cooking aflata in the preparation of kenkey (Ghana). It has been estimated that well over half of all agricultural work and food production in developing countries is done by women.

date palm as well as from sugar cane. Spirits are distillates obtained from either beer or wine and have a much higher alcohol content.

Beer

In many parts of the tropics beer is made by the sour fermentation of maize or sorghum. An example is the Nigerian pito beer (Burukutu) the

Fig. 14.6. Kenkey, a staple food of Ghana. There are several types made from fermented maize and wrapped in different kinds of leaves. Ga kenkey (let), Akporhe (right).

preparation of which varies a great deal. First the grain is soaked and allowed to germinate. It is then dried, often by exposing it to the sun (Fig 14.9). The resulting malt is ground with water and mixed with ground and boiled grain in the proportion of 1 : 1. It is then fermented for one to five days (wild fermentation). Sometimes the mash is boiled at some early stage during fermentation. While dispelling alcohol, this would help destroy harmful microorganisms which might have contaminated the original grain (See Aflatoxins p. 95). When suitable sorghum is used the grain provides the tannins which are added through the hops in Western beer. The tannins provide the slight bitterness which is desirable in beer.

In the manufacture of Western-type beer, barley is generally used. The grain is first dried and cleaned. It is then stored for about three weeks. Because of grain dormancy it cannot be made to germinate immediately. The barley is now steeped in one to three changes of water, drained and allowed to germinate for about 10 days either on the malting floor or in large containers. The sprouted grain is now dried and referred to as malt. During malting the cell walls are broken down and enzymes are liberated. These are able to degrade the starch to fermentable sugars during the

Fig. 14.7. Ashanti kenkey (Courtesy B. Bediako-Amoa.)

subsequent mashing process. In this the malt which has been ground is mixed with warm water. The starch is now broken down to sugar. The liquid 'wort' is separated and sugar and hops or hop extract are added. The liquor is now boiled, cooked and fermented for several days with the yeast *Saccharomyces cerevisiae* to produce ale or with *S. carlsbergiensis* to make lager (pure culture fermentation). Additionally, lager is aged for about three months. After fermentation is complete the liquid is sterilised, or filtered and finally bottled (Fig. 14.9a).

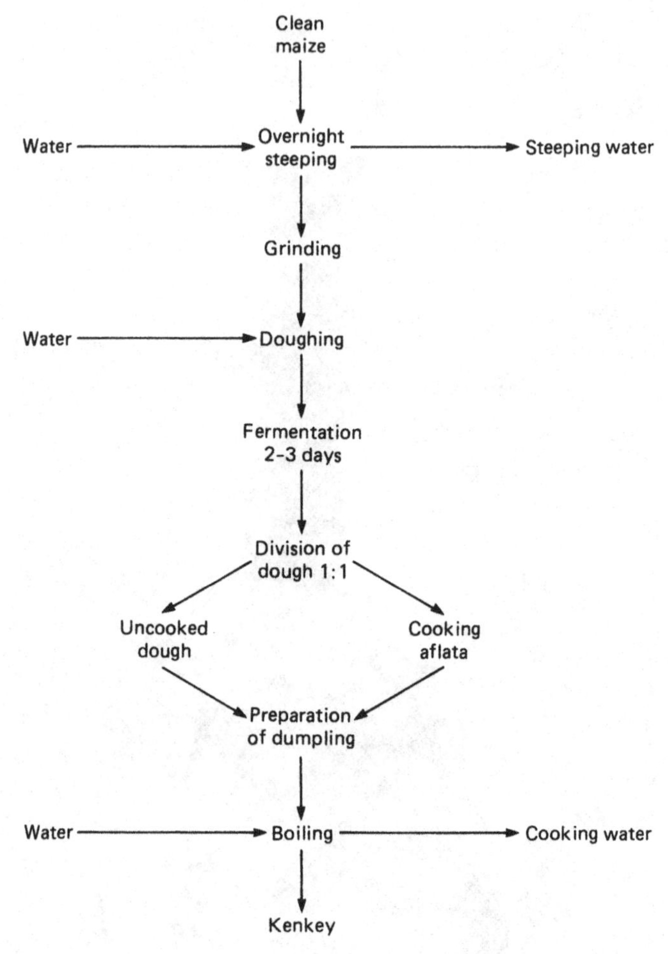

Fig. 14.8. Flow diagram of the manufacture of kenkey.

Low-alcohol beer

In the Muslim world alcohol is not permitted. Car drivers and the health conscious also often avoid alcohol and therefore an expanding market is predicted for low or 1 alcoholic beer. It is true that in many fermentations e.g. those of bread, kissra and some chapati alcohol is produced but only a trace (less than 0.05%) remains in the final product.

Beers can be divided into three groups:

(1) 'Normal' beer containing 3–6% alcohol (w/v).
(2) 'Reduced-alcohol' beer containing about 1.5% alcohol.
(3) 'Alcohol-free' beers. The alcohol content is less than 0.05%.

Fig. 14.9. Sun drying of sorghum grain in the production of pito beer (Nigeria).

Fig. 14.9a. A bottling plant at the Nigerian brewery, Aba. (Courtesy Unilever plc.)

The latter are permitted in Muslim countries and are often sold worldwide in motorway cafes.

The following are the main methods of production.

Reduced-alcohol beer

(a) Fermentation of the wort with *Saccharomyces ludwigii* gives 0.4–0.5% alcohol. It is difficult to maintain the culture and flavour is poor.
(b) The normal fermentation is checked by cooling when the alcohol concentration is 2.5%. Pasteurisation (Chapter 15) is required.

'Alcohol-free' beer

(a) The cold contact method. A large amount of yeast is added to the wort at −5°C. Under these conditions the yeast removes the wort flavour and produces the beer flavour. Only a trace of alcohol is obtained (less than 0.05%).
(b) Alcohol is removed from 'normal beer' by vacuum distillation at 52°C; 0.05% alcohol remains.
(c) Using a falling-film evaporator alcohol content is reduced to 1% in one pass. Repeated passes reduce it to 0·06%. This is perhaps the commonest method because 95.6% pure alcohol is a byproduct.

Other methods make use of dialysis (3.5% alcohol) and reverse osmosis. (Alcohol concentrations vary with the number of passes. Pronounced flavour loss occurs.)

Wine

A most important wine in many tropical countries is palm wine. This is an alcoholic beverage obtained from the fermented sap of oil or rafia palm. The former may be tapped at the inflorescence or at the standing or felled stem. When palm wine is examined under the microscope a large number of bacteria and yeast can be found. Particularly the yeasts *Candida* and *Saccharomyces* are involved in fermentation. Since fermentation is 'wild' the quality of the palm wine varies a great deal.

In Western countries, wine is the product obtained by the fermentation of grape juice. Nearly all wines are obtained from the vine, *Vitis vinifera*, which comprises both black and white varieties. Grapes require a reasonably temperate climate. The most northerly European wines come from the banks of the Rhine, the most southerly ones from Morocco. Northern wines are lower in alcohol and higher in acid content than southern ones.

To make wine the grapes are first crushed. The crushed material is transferred quickly to vats and there fermented, usually by the yeast *Saccharomyces ellipsoideus* until the alcohol content reaches 7–12%. Temperature control at 27–29°C is important. Red grapes are often fermented with their stalks which increases the tannin content. It is common practice to use sulphur dioxide for sterilisation to kill wild strains of yeast, before the introduction of the pure culture. After fermentation the wine is clarified with isinglass, charcoal, clays or pyrolidone and then bottled and matured.

Sparkling wine (e.g. Champage, Sekt) is made by adding a small quantity of sugar and pure yeast culture to the bottled wine. It is then kept for six to eight weeks at 16–19°C. During the fermentation the yeast rises in the bottle and adheres to the cork. When fermentation is complete, the neck of the bottle is frozen and the yeast and the cork removed together. A new cork is then inserted and fastened firmly with wire.

Spirits

Distillation and the resulting higher alcohol content of spirits have two consequences. The bulk of water has been reduced, so reducing transport costs and the keeping properties have also been improved. At the end of fermentation, the brew, be it wine or beer, is placed into a distillation apparatus such as a pot still for batch production. This consists of an hemispherical copper pot with a cover shaped into a tube. This tube passes into the condenser. In principle, this is the method used for the distillation of, for example, palm wine in most tropical countries. Alternatively, the 'Coffey still' can be employed which is a continuous steam heater.

Spirits, like wine, require maturing. While wine is normally matured in bottles, spirits are matured in wooden casks. Brandy is made by the distillation of wine and whiskey from fermented grain such as barley, rye or maize. Western gin is basically grain alcohol containing several spices. The relatively limited storage life of gin is due to the oxidation of oils which may cause rancidity or bitter flavours. The West African 'illicit' gin (known by various names in different areas, e.g. Kiki, 'push-push', Ekpeteshi) is made by the distillation of palm wine which has been fermented for about four days. Liqueurs are neutral spirits with an infusion of various herbs and are sweetened with sugar syrup.

The most important constituent of all fermented beverages is alcohol. It is lowest in beer and stout being of the order of 2–4% although strong ale may contain up to 6% alcohol. Wines usually contain 8–10% whilst fortified wines such as sherry, port and madeira may be strengthened with additional alcohol to a level of 18%. Spirits such as brandy, whiskey and liqueurs usually contain 40% alcohol. Often the term 'proof spirit' appears on the label. This is defined as 'spirit which at 51°F weighs exactly the twelve-thirteenth part of an equal measure of distilled water'. It is in fact a mixture of almost equal parts of absolute alcohol and water.

The effect of alcohol is entirely depressant. First alcohol depresses the inhibitory centres and soon balance is affected. Intoxication has been divided into four stages easily remembered as dizzy and delightful, drunk and disorderly, dead drunk and danger of death. These stages correspond roughly to a blood alcohol content of 1, 2, 3 and 4 mg per ml of blood.

Industrial fermentations

As the earth's resources are slowly being depleted, alternative renewable resources are being explored. A new expression has been coined to cover this field: Biotechnology. This has been defined as the application of scientific and engineering principles to the processing of materials by biological agents to provide goods and services. Its practice is thousands of years old. In the West wheat, rye and barley have been fermented into bread and beer, grape juice into wine and milk into cheese and yoghurt. In the East biotechnology has fermented soya beans into soy sauce, tempeh and miso, and rice into saké. In Africa sorghum, maize and millet have been fermented into sour beers, porridges and dumplings, palm juice into wine, and cassava has been detoxicated.

Industrial fermentations date back to the turn of the century when ethanol, butanol and acetone were produced. World War II saw the production of antibiotics and later came steroids, enzymes and some vitamins. In the early 1970s biotechnology received a revolutionary impulse through gene manipulation where genetic material from a donor cell was transferred to the genetic structure of a host. This was called recombinant DNA research, popularly called 'genetic engineering'. It is therefore necessary to consider the effect on tropical countries of two separate techniques: chemical fementation technology and genetic engineering.

The first has already been considered on pp. 230–243 in connection with traditional fermentations. However some of these have been scaled up and used on an industrial scale. The amounts of chemicals produced

Table 14.2 *World production of some chemicals produced by industrial fermentation (in 1000 tons)*

Antibiotics	23	Glutamic acid	200
Single cell protein	30	Lactic acid	15
Citric acid	300		

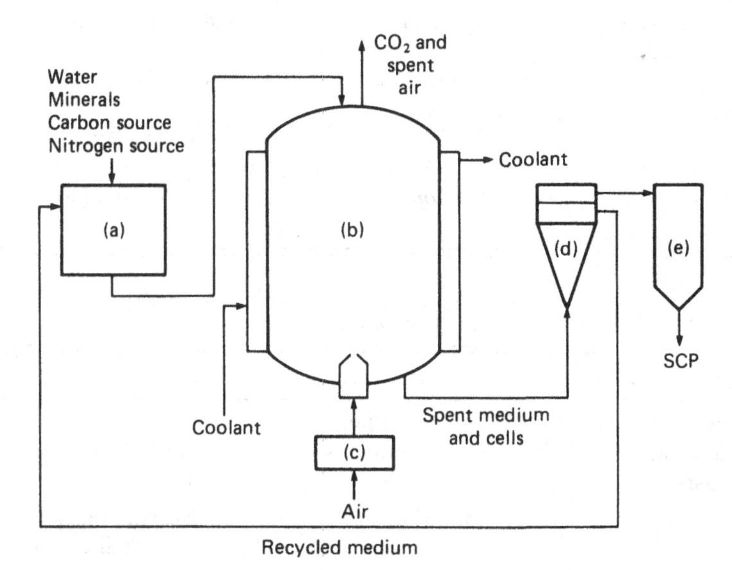

Fig. 14.10. Continuous fermentation plant for the production of single cell protein (SCP). (a) Nutrient tank, (b) fermenter, (c) air filter, (d) centrifuge, (e) drier.

worldwide at present by industrial fermentation are given in Table 14.2.

Most industrial fermentations require very pure strains of microorganisms. Therefore the growth medium, the air and the equipment must be kept sterile. Most of these fermentations require a pH of 7 and this is favourable for the growth of many contaminating microorganisms. Some processes only take place efficiently either in the complete absence of air (if the organism is anaerobic) or require large volumes of sterile air (if the organism is aerobic). Figure 14.10 shows the diagram of a continuous fermentation plant for the production of single-cell protein. Sterile water and nutrients as well as a carbon and nitrogen source are passed into the fermenter together with sterile air. Spent air and carbon dioxide are removed and usually the system must be cooled. The spent medium and the microorganisms are passed to a centrifuge where the liquid is separated from the cells. The latter are then dried. The medium can be

Table 14.3 *Production by fermentation of some products used in the food industry*

Substrate	Organism	Product	Food use
Molasses, starch hydrolysate	*Saccharomyces*	Ethanol	Alcoholic beverages
Ethanol	*Acetobacter*	Acetic acid	Vinegar
Molasses, starch cellulose	*Aspergillus niger*	Citric acid	Soft drinks
Molases, starch, hydrolysate, whey	*Lactobacillus delbruckii*	Lactic acid	Flavour, acidulant
Starch hydrolysate	*Xanthomonas campestris*	Xanthan gum	Thickener
Methanol	*Pseudomonas*	Single cell protein	Food additives

Table 14.4 *Systems used in biotechnology*

Experimental system	Biological process	Method
Microorganisms i.e. bacteria, moulds or yeasts	Fermentation:	
	– continuous	Fermenter
	– batch	Vat
Enzymes	Biocatalysis	Enzyme reactor (immobilised enzymes)
Plant and animal cells	Growth *in vivo*/*in vitro*	Cell or tissue culture

recirculated and if it contains desirable fermentation products these can be separated further.

Examples of processes where the microorganism itself is harvested for food are the production of the bacterium *Pseudomonas* from methanol, of the yeast *Candida* from crude gas oil, of the filamentous fungus *Aspergillus* from carob bean waste and of the alga *Spirulina* from carbon dioxide. (Being a green plant, *Spirulina* is an autotroph.)

An important example of the anaerobic type of fermentation is the production of acetone–butanol. Since fermentation usually takes place in a large volume of water, the waste disposal can be a problem. Fermentation processes require a high capital investment. Some of them yield products used in the food industry and are listed in Table 14.3.

In developed countries a great deal of experimentation is carried out (Table 14.4) and Fig. 14.11 shows a series of small-scale fermenters which are used to grow microorganisms in the laboratory.

It is also possible to enrich low-protein foods by fermentation in the

Fig. 14.11. A set of laboratory-scale fermenters.

solid state rather than in a liquid medium. For example *Aspergillus niger* can be grown aerobically on cassava meal which has been steamed for 10 min. The meal is inoculated with ammonium salts and urea as nitrogen sources and the spores of the fungus. After 24 h at 50–55% moisture and at 35°C the protein content of the meal has been doubled.

Another important branch of biotechnology is the use of immobilised enzymes (p. 23).

The advantages of biotechnology for the tropics are almost entirely in the field of classical fermentation. Wild fermentation must be replaced by pure cultures. The protein content of carbohydrate foods like cereals and tubers can be increased using suitable mircoorganisms. Anti-nutrients can be removed and toxins (cyanide, aflatoxin) destroyed. There is a huge amount of biomass in the wet tropics although it has never been quantified. Sometimes land has had to be withdrawn from food production to produce biomass for energy (wood, ethanol) because many foods cannot be eaten raw. Nevertheless there are large amounts of cellulose byproducts which can be converted. This process is slow but more effective microbial strains are being developed. Single-cell protein is readily produced in fermenters when water is scarce and cassava chips have been treated with a nitrogen source and relevant fungi to increase protein content. Seeds can be treated with freeze-dried nitrogen-fixing microorganisms and blue-green algae are used to fix nitrogen in rice paddies. As regards drinking water, oil spillages can be removed and heavy metal contamination, as for example in Sierra Leone, can be absorbed although suitable microorganisms are at present not very effective.

What about the disadvantages of biotechnology for the tropics? Plant cells can now be grown initially in glass dishes and then in a fermenter. Interesting plant cells are those which yield highly priced secondary metabolites and many of those come from the tropics – wild basil, buchu, capsicum, dill, eucalyptus, guava, lemon grass and mint for flavours, avocado for sugars, cedarwood for sesquiterpenes, chinchona for quinine, tonka beans for lactones and pyrethrum for insecticide. At present saffron, capsicum, pyrethrum and cocoa have been grown sucessfully in fermenters of various sizes. A relatively cheap and consistent product produced in this way would be preferable to a variable and more expensive import from some tropical country. The consequences for the economy of some of these are obvious.

As regards genetic engineering, the method consists in removing a particular gene from one organism, producing identical copies (clones) of the gene and inserting them in working order into another species. In practice, the complex arrangement of genes in animals and plants is not well understood and therefore microorganisms are used. Of particular interest are genes producing desirable enzymes which can be inserted into bacteria or yeasts. It would be very useful if enzymes could be made cheaply and in large quantities. It will also be possible to improve the properties of such enzymes. Already it has been possible to produce thaumatin (Chapter 10) and rennin, the milk-clotting enzyme used in cheese production, by inserting the relevant gene into yeast.

Genetic engineering presents interesting possibilities for the tropics. In the host disease resistance can be increased; in the pathogen the disease-producing genes can be removed and the pathogen used as a vaccine. Malaria, hepatitis, rabies and schistosomiasis could be wiped out. However such research is exceedingly expensive and requires an extensive scientific infrastructure.

15

Food preservation

If kept beyond a certain period most foods will deteriorate, often with changes in texture, taste and smell. More important still, poisonous products may be produced. These changes can be due to microorganisms i.e. bacteria, yeasts and moulds. Essentially food preservation is based on methods that kill microorganisms or stop them from growing without making the food unsuitable for human consumption later.

The most important methods of food preservation are those based on physical treatment, namely freezing, drying, pasteurisation, canning, irradiation and those based on chemical modification or additives. The physical methods are dealt with in this chapter, the chemical ones in Chapter 16.

Blanching

Before preservation by, for example, freezing or canning vegetables are often blanched. Blanching consists of heating the vegetable briefly with either water or steam. The cooking or the destruction of microorganisms is not the primary object, but blanching destroys the enzymes which may affect colour, texture, flavour and nutritive properties during subsequent processing and storage. Blanching also destroys the semi-permeable membrane of the cells and expels gas from the tissue. This may collapse and it is easier to pack the material into cans. There is some destruction and leaching of vitamins, especially vitamin C during blanching.

The effectiveness of blanching is often assessed by measuring the destruction of the enzyme peroxidase. This is not necessarily connected with deterioration but it is used as an indicator and is easily measured. The enzyme is extracted and the brown products resulting from its action in the presence of pyrogallol are extracted with ether and the intensity of the colour measured in a colorimeter.

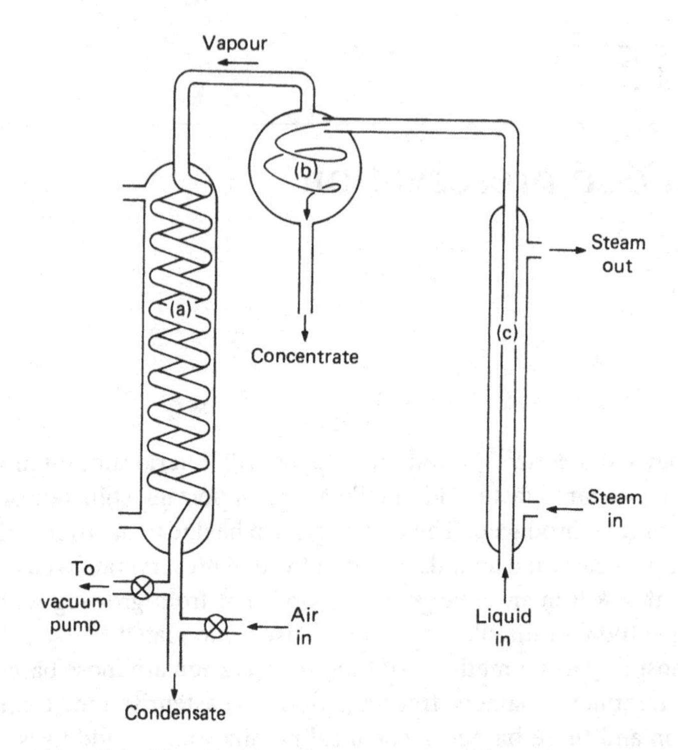

Fig. 15.1. The climbing-film evaporator: (a) condenser; (b) cyclone; (c) calandria.

Evaporation

Partial evaporation of liquids is carried out mainly for two reasons. Fruit juices are often concentrated to reduce transport costs, pasteurised (p. 266) and then bottled or frozen. They can then be diluted again just before use. Other liquid foods, e.g. milk, are partially evaporated before being subjected to final spray-drying (see below) or roller-drying (see below). Both these processes work best at 20% total solids concentration.

There are quite a number of different evaporators. The simplest is the evaporating pan, where the liquid is boiled at normal atmospheric pressure. With temperature-sensitive liquids use is made of the principle of lowering the boiling point of the liquid by reducing the pressure above. Therefore these evaporators are used under vacuum. There are tube, plate or centrifugal evaporators. A simple tubular type is the climbing film evaporator (CFE) (Figs. 15.1, 15.2).

The liquid is forced to flow as a thin film over the heated surface of the calandria being carried up the tube by the high velocity vapour stream in

Fig. 15.2. A climbing-film evaporator in a teaching laboratory.

the centre of the tube. The heat transfer to this thin film is more rapid than would be obtained by heating the liquid in bulk.

In the laboratory type shown, there is only one tube in the calandria but in a large industrial plant there can be several. The boiling liquid now passes into the cyclone and from there into the concentrate receiver. Its vapour passes out of the cyclone into the condenser and the liquid condensate is collected.

An air pump keeps the system under vacuum and so reduces the boiling point of the liquids. An airbleed allows the control of the vacuum and

Table 15.1 *Freezing point of some common foods*

Food	°C
Groundnuts	−8.5
Walnuts	−6
Banana, coconut	−4
Veal, lamb	−3
Beef, fish	−2
Lettuce, cabbage	−0.5
Water	0

thus the temperature of boiling. The airflow can be controlled either manually or by computer.

Freezing

Practically all living cells contain a great deal of water and dissolved materials such as sugars, salts and many other organic and inorganic molecules. All these depress the freezing point which is therefore lower in foods than in pure water. Table 15.1 gives the freezing point of some common foods.

Preservation by freezing is based on the fact that microorganisms will not grow below their freezing point. Most biochemical changes are also slowed down considerably by low temperatures, although some oxidation reactions can still be quite rapid. (Some frozen foods with high fat contents can become rancid in about two weeks.) The type and speed of freezing largely determine the subsequent texture of the food. The rate of freezing depends on how quickly the 'freezing front' penetrates into the interior of the food. As the tissue freezes ice crystals are formed which may puncture the cells. The slower the freezing process, the larger will be the ice crystals. The larger these are, the greater is the change in the tissue, the greater the change in texture on thawing and the greater the drip loss.

Damage during freezing is also commonly caused by osmotic pressure differences due to the change in concentration of solutes as freezing proceeds.

A simple mechanical refrigerator (Fig. 15.3) consists of an evaporator, compressor, condenser and expansion valve, filled with a slow-boiling liquid or its vapour. The heat absorbed from the cold room or the product results in boiling of the refrigerant in the evaporator. The vapour is then

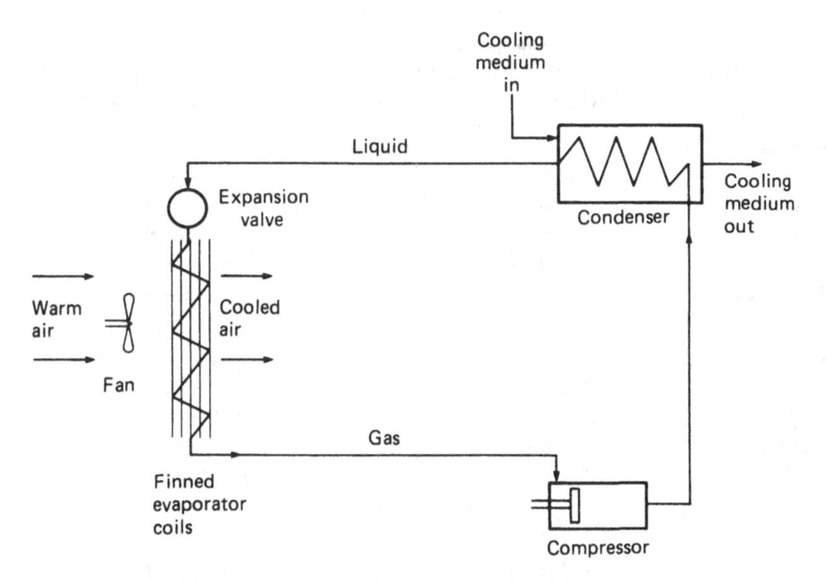

Fig. 15.3. Diagram of a refrigeration unit.

removed by the compressor, which is normally a reciprocating pump. Pressure is now increased and heat removed in the condenser. High-pressure liquid passes through the expansion valve where the pressure of the liquid is reduced before being piped into the evaporator.

A simple method of freezing a product is by immersing it directly into a bath or spray of refrigerant such as glycerol, but to avoid contamination by the refrigerant the product must be covered. Next in efficiency is the plate freezer. Here the product is squeezed between two refrigerated plates, one above and one below the product. In the blast freezer a cold airstream is blown over the product. In a cold-room there is little air movement. Hence freezing in a cold-room is very inefficient.

Some foods, such as avocado, banana and some citrus fruits are easily damaged by low temperatures. Others, such as some berries and leaf vegetables are fairly unaffected even after relatively long cold storage.

If the food is of large size, conduction of heat through the food is slowed down and this reduces the freezing rate at the centre.

Some non-living foods, such as meat, fish and dairy products which are easily affected by their own enzyme systems and by microorganisms are often stored frozen between −18 and −35°C. Living foods such as fruit and vegetables are still metabolically active and have a natural resistance to deterioration. They are often stored chilled. Chilled foods are perishable foods or food products which, to remain wholesome, need to be kept at temperatures between −1 and +8°C and which are not frozen.

Table 15.2 *Domestic safe storage times*

Item	Refrigerator (below +5 °C)	Freezer (−18 °C)
Raw meat		
Joint	3 days	3–8 months
small pieces	2–3 days	3–8 months
Minced	1 day	3–4 months
Offal	1–2 days	3–4 months
Bush meat	3–4 days	6–8 months
Fish		
Raw	1 day	3–6 months
Cooked	2 days	3–4 months
Frozen	2 days	3–4 months
Shellfish	use same day	1 month
Smoked	2 days	6–12 months
Fruit		
Soft	2–3 days	6–9 months
Stonefruit	3–7 days	6–9 months
Citrus	14 days	3–6 months
Pineapple	10 days	Not recommended
Vegetables		
Greens	5–7 days	6–9 months
Herbs	6 days	6–9 months
Green beans	1 day	6–9 months
Tomatoes	7–14 days	Pulped 6 months
Dairy products		
Hard cheese	1–4 weeks	4–6 months
Soft cheese	1 week	1 month
Ice cream	1 day in ice compartment	3 months
Milk, cream	3–4 days	Not recommended

Table 15.2 gives the safe storage times for various foods under domestic conditions in a refrigerator and freezer. Table 15.3 gives industrial recommended storage conditions where temperature and relative humidity can be controlled.

During air-freezing, water may be withdrawn from the tissue by sublimation. Although freeze-drying is based on this (see below) it is undesirable on normal freezing and is referred to as 'freezer burn'. It may alter irreversibly colour, texture, flavour and nutrient value of frozen foods. Roast beef for instance acquires the appearance of light brown paper due to freezer burn. The effect can be avoided by proper packaging.

In the developed world the amount of frozen food consumed has

Table 15.3 *Industrial recommended storage conditions*

Item	Temperature (°C)	Relative humidity (%)	Maximum safe storage time
Raw meat			
Beef	−2–0	90	5 weeks
Lamb	−1–0	90–95	2 weeks
Pork	−2–0	90–95	1–2 weeks
Offal	−1–0	85–90	1 week
Eggs	−1.5–0	85–90	6–7 months
Fish			
Raw	0.5–4	90–95	3 weeks
Smoked	5–10	50	6–8 months
Salted	5–10	90–95	10–12 months
Frozen	−23–−17	90–95	8–10 months
Fruit			
Avocado	5–10	90	2–4 weeks
Lemon (ripe)	0–2	85–90	4–6 weeks
Orange	0–4	85–90	2–4 months
Guava	7–10	85–90	3 weeks
Mango	8–10	85–90	3–6 weeks
Passion fruit	6–7	80–85	4–5 weeks
Pineapple	5–7	90	2–3 weeks
Pawpaw	4–5.5	85–90	4 weeks
Vegetables			
Greens	0–1	90	2 weeks
Herbs	0–1	85–90	1–2 months
Green beans	−0.5–0	85–90	2–3 weeks
Tomatoes (ripe)	2–4	85–90	2–4 weeks
Dairy products			
Milk, cream	0–2	—	3–4 days

increased every year since 1980. In the United States more than half of all married women hold jobs. Two-income families have more money but less time for food preparation. Hence frozen foods and microwave ovens are welcome.

Frozen or chilled foods must be kept uninterruptedly at the recommended temperature from producer to distributor, to retailer and finally to the consumer.

In developing countries small shops and consumers generally lack refrigerators. The housewife may shop every day for fresh meat and vegetables. For this reason chilled or frozen food is not so important in many tropical countries.

Drying

Dehydration in air

Drying in the early stages takes place at a rate governed entirely by surface evaporation and the rate is given by

$$hA(\theta_a - \theta_w)$$

where h is the heat transfer coefficient from the air to the food surface, A is the exposed of the food, θ_a is the air temperature and θ_w is the wet bulb temperature of the air. (This is tabulated for different air relative humidities.) Obviously heat must go into the product for moisture to come out. The coefficient h is about 5 to 25 $W\,m^{-2}\,K^{-1}$ (Watts divided by square metres multiplied by the absolute temperature) and depends on air velocity (e.g. in sun-drying, how windy conditions are). The system governed by the equation is in essence similar to water evaporating in a dish. This type of drying will occur until the moisture content of the food reaches about 200% on a dry weight basis.

From this moisture content down, the rate is controlled by the diffusion of moisture from within the food to its surface. There are theoretical equations available for predicting drying rates in this process for definite shapes of food (e.g. cylinders, spheres, slices). For a slice, the equation has a solution giving

$$\ln\left(\frac{W - W_i}{W_i - W_e}\right) = \ln\frac{8}{\Pi^2} - \frac{D\Pi^2 t}{L^2}$$

where W is the moisture content after time t, W_i is the initial moisture content, W_e is the equilibrium moisture content (which depends on the relative humidity of the air, but for a fruit in low humidity air will be around 5% on a dry weight basis) and L is the slice thickness. D is the diffusion coefficient and this is very difficult to predict. It is basically a measure of the ability of one material (in this case water) to diffuse through another (the food). It depends largely on the changes in the food (shrinkage, chemical, and particularly surface change) during drying. It is determined in the laboratory where one can dry under strictly controlled conditions. The value of D often lies between 1×10^{-7} and $1 \times 10^{-9} m^2 s^{-1}$. The equation is best used for predicting the effect of a change in drying conditions (e.g. changing thickness, air temperature or humidity) rather than in obtaining *absolute* drying rates.

Perhaps the simplest dehydration process and one common in the tropics is sun-drying (Figs. 9.7 and 14.9). Its principal advantage is

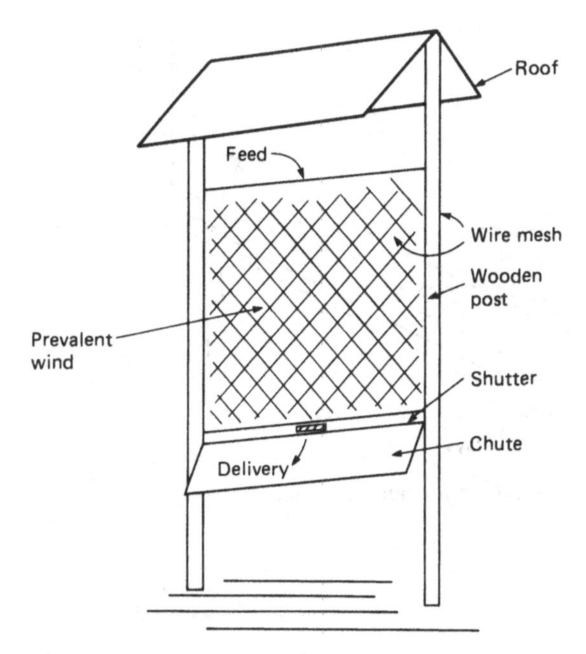

Fig. 15.4. Wind-drier for cassava chips.

that it is very cheap. Disadvantages are that it is seasonal, not very sanitary, requires a relatively large area and takes relatively long, particularly with high relative humidity or regular rain. As a result total solids are lost through tissue respiration and fermentation. The properties of the final product are also greatly altered through enzymatic or biochemical agency. Compare, for instance, grapes with raisins, plums with prunes and fresh yam or fish with the sun-dried product.

Another arrangement is the wind drier shown in Fig. 15.4. Here maize cobs, potato, yam or cassava chips are sandwiched between two vertical wire meshes. The drier is fed at the top and unloaded over the shute by opening the slide. Frame and roof are made of wood. Delicate materials like fruit cannot be dried because the pressure at the bottom of the drier is too high. These should be dried in stands with horizontal wire netting. The relative humidity of the air should be below 60% and wind velocity above 4 m/s.

A sun-drier suitable for many seeds, fruits and vegetables is shown in Fig. 15.5. The wooden trunking is arranged at 10–12° facing the sun. The food to be dried is placed on wire trays covered with UV-light resistant plastic sheet. The hot air is drawn downwards through the material to be dried and escapes through the air duct. This should be painted black to increase the draught. Depending on air resistance the material depth

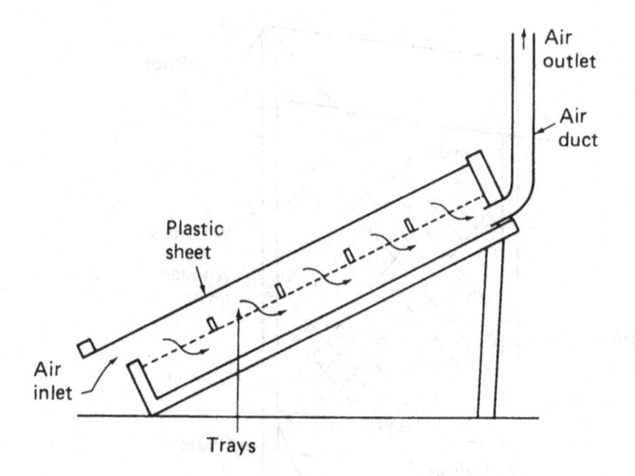

Fig. 15.5. Sun-drier for bulk materials.

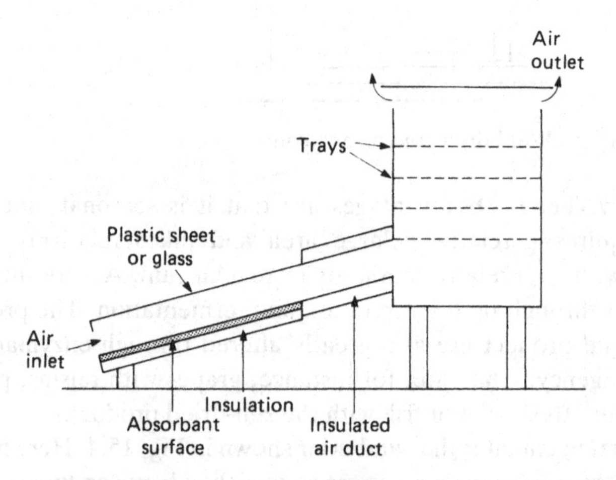

Fig. 15.6. Indirect sun-drier for bulk materials. Usually sunlight is one of the great assets of the tropics. There are solar research stations, e.g. in Bamako (Mali), and Abidjan (Ivory Coast).

should vary between 3 and 10 cm. Temperatures, which may reach 60–70°C, must be checked to avoid overheating.

This system can also be used for indirect drying. A solar collector is constructed from a heat-absorbent surface (e.g. metal sheet painted black) with an insulating layer (wood, expanded polystyrene) underneath, and glass or UV-light resistant plastic sheet on top. This is connected to the drier (Fig. 15.6).

Modern industrial driers are based on the same principle. There is a heat source and air movement is achieved by powerful fans which drive

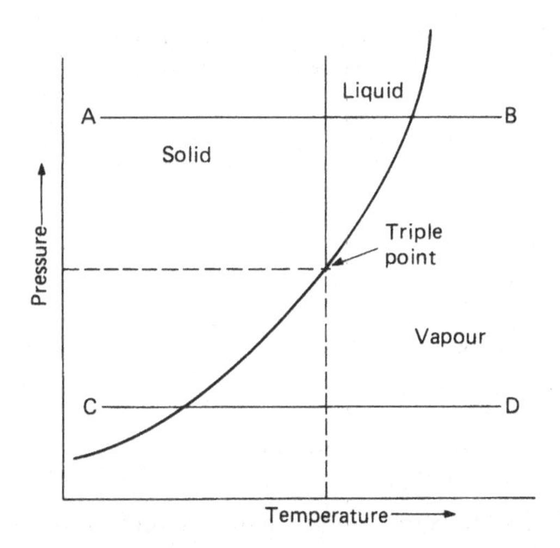

Fig. 15.7. A phase diagram. The 'triple point' of a substance is the point at which all three phases (solid, liquid and vapour) coexist in equilibrium.

the air through the product. A modification is the tunnel drier. Here the product is placed on a conveyor belt which passes through a hot-air tunnel. This process is often used for vegetables.

Freeze-drying

When the temperature of water is reduced to 0°C under atmospheric pressure it freezes i.e. turns into solid ice. If water is heated it turns into steam i.e. it vapourises (line A–B in Fig. 15.7). If the pressure is reduced sufficiently, ice will vapourise without passing through the liquid phase (line C–D in Fig. 15.7), this process is called sublimation. If water in the food were present in the pure state, it would be possible to freeze-dry it at 0°C and a pressure of 4.6 mm mercury. However since the water exists in solution or in a combined form most freeze-drying takes place at −10° to −40°C.

For freeze-drying on an industrial scale the food is placed on trays or belts in a sealed chamber. The temperature and pressure are reduced and the product frozen. When a sufficient vacuum has been reached, the frozen food is carefully heated by hot-water coils, electric resistance or infrared sources so that sublimation is maintained without any ice melting within the food. A diagram of a freeze-drying unit is given in Fig. 15.8.

This process is used with solids such as fruit, vegetables, meat or shellfish as well as with liquids including milk, fruit juice or coffee extract.

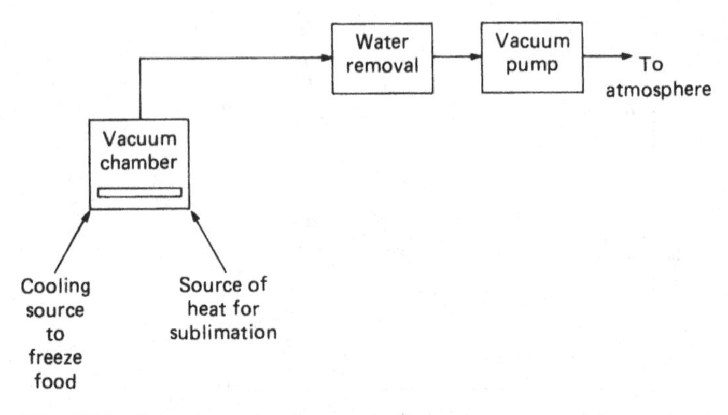

Fig. 15.8. Diagram of a freeze-drying unit.

The dehydrated food is very porous and light and reconstitutes very quickly with water to give an excellent final product. Unfortunately the dehydrated material is hygroscopic and must be packed in sealed containers, often under vacuum, or in an inert gas atmosphere. The cost of treatment is also very high and it is only used where a very high quality product is essential. Fig. 15.9 shows a freeze-drying chamber.

Roller drying

If a solution, slurry or thin paste is spread on a heated rotating roller, the water will evaporate leaving a dry film on the roller. This may be scraped off. This method is used for drying milk, porridges, mashed tubers as well as fruit and vegetable products. Roller-dried materials undergo fairly intense heat treatment. This means that bacteriological properties are good, but the food is changed more than e.g. spray-dried material (see below). For instance the proportion of milk powder which dissolves on reconstitution is lower with roller- than with spray-dried milk. Flavour and colour may also be affected. With cereals, a desirable roasted flavour is produced. The starch is also gelatinised during the heating process and therefore the product does not need further cooking after reconstitution. Vitamins may suffer during the process and roller-dried baby foods are often enriched with vitamin D and roller-dried English potato with vitamin C. With juices roller-drying is often preceded by some preliminary concentration process (e.g. evaporation) because roller driers (as well as spray driers) work best at a total solids concentration of about 20% (Figs. 15.10–15.13).

Fig. 15.9. Vacuum chamber for industrial freeze-drying. (Courtesy Nestlé.)

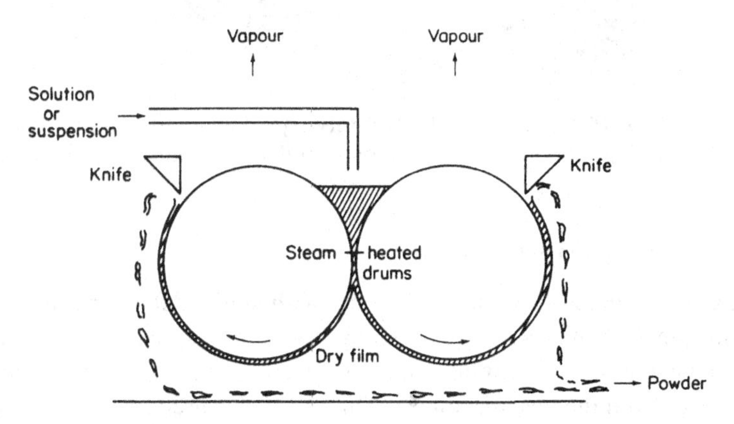

Fig. 15.10. Diagram of a simple roller drier.

Fig. 15.11. This small laboratory type double-drum roller drier is being used for the production of roller-dried milk.

Spray drying

This is a method of drying liquid foods with minimal flavour change. The principle is to mix a very fine spray with hot gas, usually air, thereby achieving almost instantaneous drying and then to separate the dry product from the gaseous phase. This method is used for the production of milk and egg powders, instant coffee, coconut milk, and other heat-

Fig. 15.12. Industrial single-cylinder roller drier. (Courtesy R. Simon and Sons.)

sensitive materials. It is a rather expensive process but under the right conditions gives a very high quality product with little loss in nutritive properties and excellent reconstitution.

The diagram Fig. 15.14 shows the layout of a typical spray drier. The liquid is fed onto a cone which, driven by an air turbine, rotates very fast inside the drying chamber. Hot air is also passed into this chamber and the droplet spray from the cone is dried almost instantly into fine globules. The powder is separated from the hot, moist air in the cyclone and falls into the powder hopper. (A cyclone is a chamber in which a suspended powder is separated from a gas by centrifugal force.)

Fig. 15.13. A battery of industrial roller driers. (Courtesy Nestlé.)

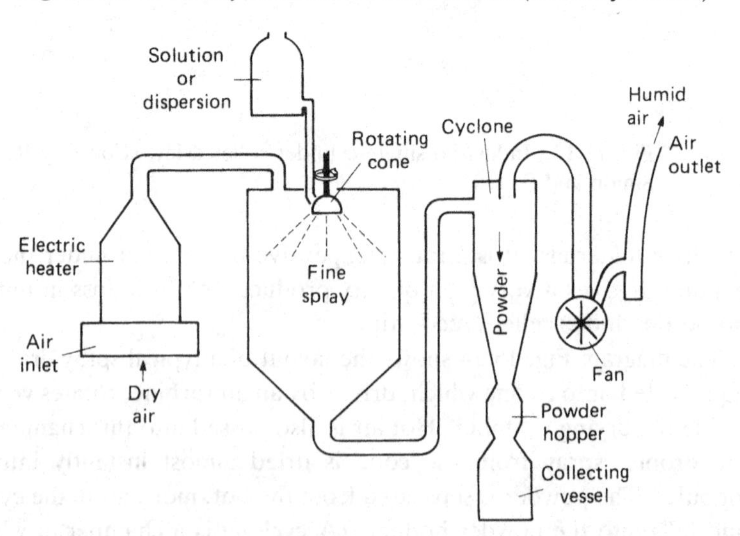

Fig. 15.14. Diagram of the small laboratory-scale spray drier shown in Fig. 15.15.

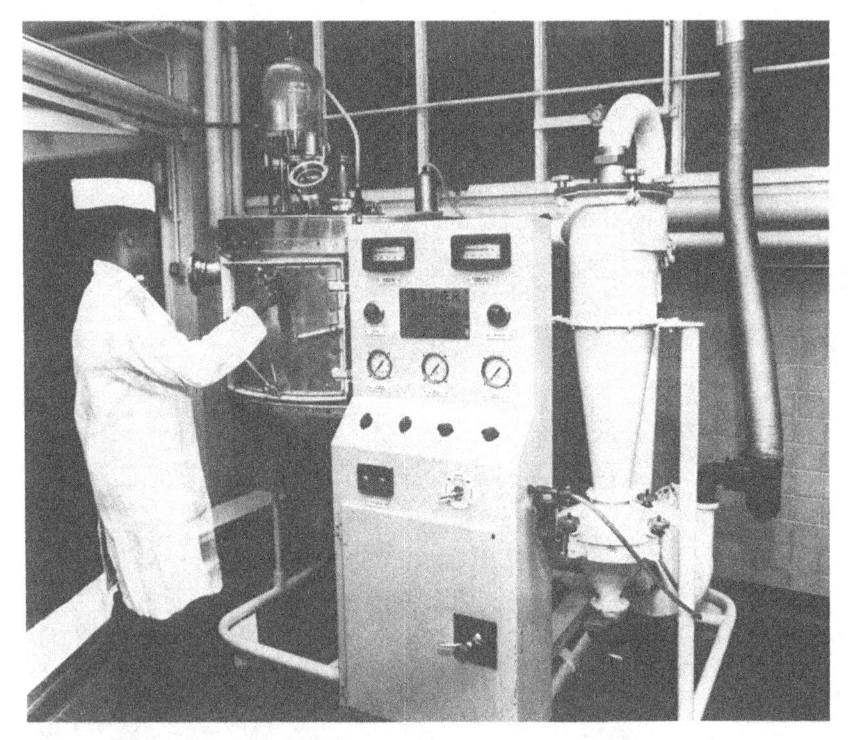

Fig. 15.15. A small laboratory spray drier (see text).

Figure 15.15 shows a small laboratory-type spray drier. The glass bottle with the liquid feed can be seen above the steel drying chamber on the left. On the right are the cyclone, powder hopper and collecting vessel. Behind is the fan with the air outlet pipe to the roof of the building. Heater and air intake lie behind the drying chamber and are not visible in the picture. Figure 15.16 shows the lower section of a large industrial spray drier.

Spray-dried and roller-dried powders are easily distinguished under the microscope. Figure 15.17 shows spray-dried milk powder which appears as small granules. Roller-dried powder by contrast, appears as small jagged plates (Fig. 15.18).

Pasteurisation and canning

Both pasteurisation and canning subject the food to wet heat with the object of greatly reducing or destroying microorganisms and so allow the food to be stored for some time without deterioration.

Fig. 15.16. Lower section of an industrial spray drier for instant whole milk. (Courtesy Stork Amsterdam International Ltd.)

Pasteurisation

On pasteurisation most microorganisms in liquids are destroyed although spores generally survive. Fruit juices, alcoholic drinks and egg products are frequently pasteurised, but perhaps the most important function of pasteurisation is the destruction of the tubercle bacillus (*Microbacterium bovis*) in milk. Spore formers usually survive the process and may grow later. Therefore pasteurisation does not give unlimited protection and is often used together with some other method of preservation. These may be refrigeration with egg or milk products, anaerobic storage with beer, maintenance of an acid medium with pickles and fruit juices together with high sugar concentration or other chemical treatment.

The time and temperature of the treatment depend to a large extent on

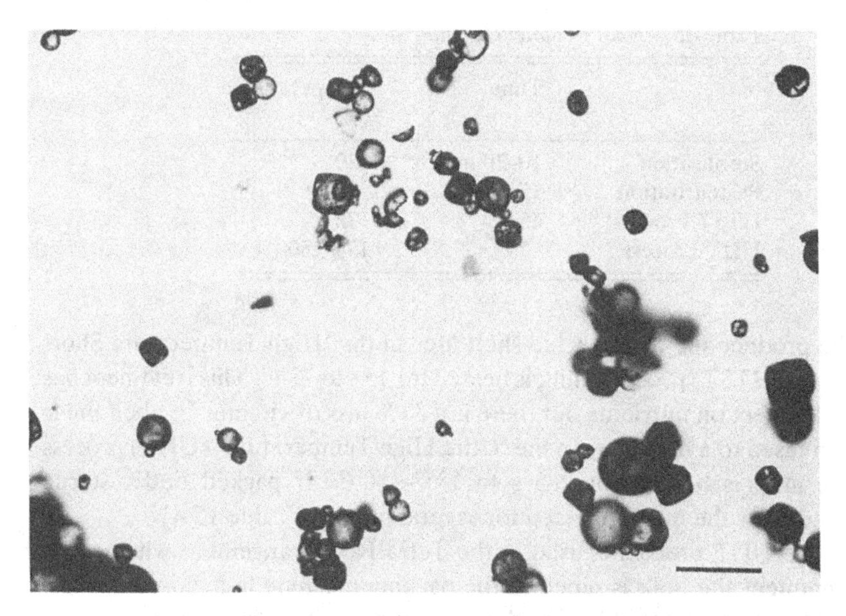

Fig. 15.17. Spray-dried milk powder as seen under the microscope. The particles are not necessarily spherical but always have a rounded outline. Bar = 100 μm.

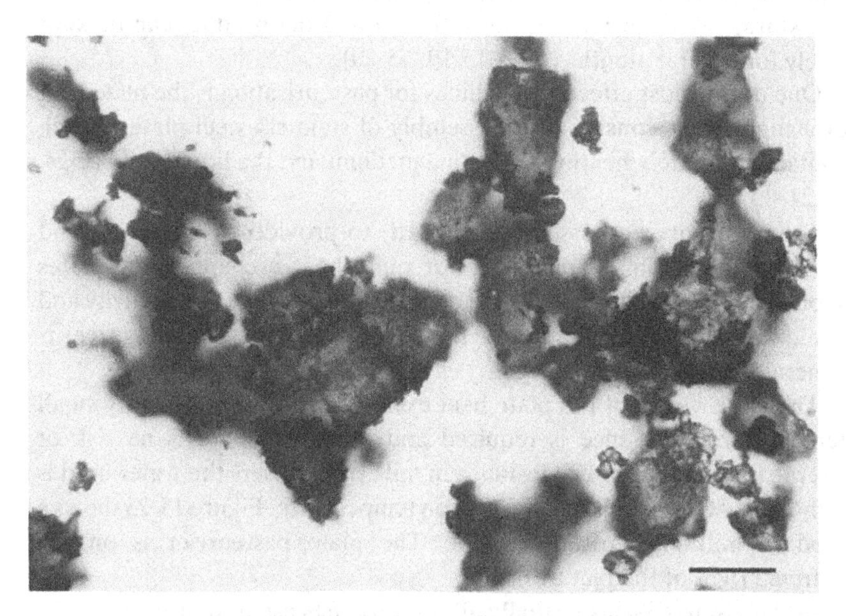

Fig. 15.18. Roller-dried milk powder showing flakey structure. The particles have a jagged outline and are usually larger than spray-dried particles unless they have been sifted. Bar = 100 μm.

Table 15.4 *Heat treatment of milk*

	Time	Temperature (°C)
Sterilisation	10–20 min	120
Pasteurisation	15–20 s	72–76
HTST Process	15 s	72
UHT Process	2–5 s	135–150

the product and its expected shelf-life. In the 'High Temperature Short Time' (HTST) process, milk is heated for 15 s to 72°C. This treatment has little effect on nutrients but there is a 20% loss of vitamin C. Shelf life is increased to a few days. In the 'Ultra High Temperature' (UHT) process the milk is heated for 2–5 s to 135°–150°C. If packed under sterile conditions the milk will keep for several months (Table 15.4).

The UHT process is used in the Tetra Pak arrangement where after treatment the milk is piped to the packing machine in a closed system. The packing material is passed through a bath of hydrogen peroxide and any adhering liquid is removed by sterile hot air. This aseptic filling system has allowed fundamental changes in the handling of perishable liquids such as milk, coffee, tea, soups and fruit juices. On distribution and storage, refrigeration is not required and the product can be kept safely for several months (Figs. 15.19, 15.20).

One of the most effective machines for pasteurisation is the plate heat exchanger. This consists of an assembly of stainless steel plates, which contain alternately heating or cooling medium and the liquid food (Figs. 15.21–15.23).

The plates are embossed (corrugated) to provide both strength and turbulent flow. This is necessary for rapid heat exchange. The plates should also be as thin as possible, should have good heat conductivity and avoid metallic contamination of the food. Therefore, stainless steel is generally used.

The great virtue of the plate heat exchanger is that only a very small temperature difference is required and therefore there is no risk of overheating the food. For instance in milk pasteurisers the water used is only 2° or 3°C above the pasteurisation temperature. Figure 15.23 shows a modern milk-pasteurisation plant. The plate pasteuriser is on the extreme right of the picture.

In the United States virtually all egg products are pasteurised at 60°C for 3.5 min to kill *Salmonella*. In the United Kingdom pasteurisation for 2.5 min at 64.5°C is practised. The Americans feel that the British

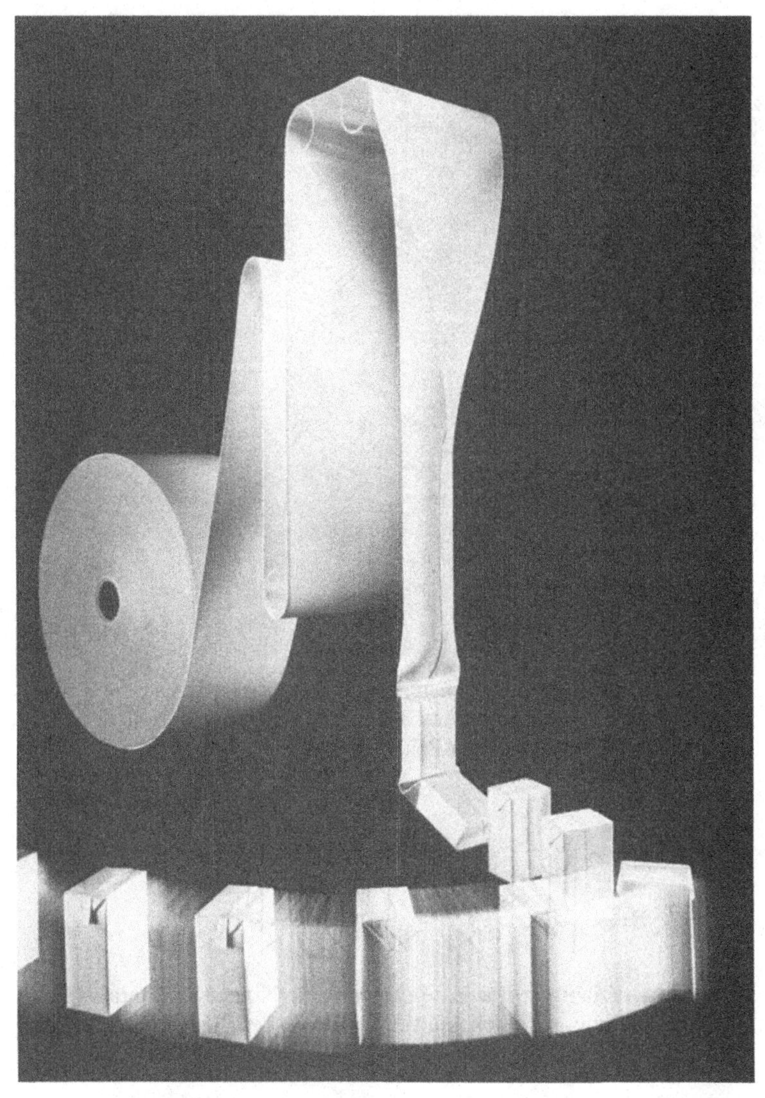

Fig. 15.19. The basic concept of the Tetra Pak is to form a tube from a roll of plastic-coated paper, fill it under aseptic conditions and seal it below the liquid level. This avoids an air space above the liquid. (Courtesy Tetra Pak.)

temperature is too high, and harms technological properties (e.g. cake volume), but this higher temperature allows the very convenient amylase test. The enzyme amylase contained in the egg is denatured under UK but not under US conditions. Hence there is no suitable test for pasteurisation in the United States. Pasteurisation must be carried out very

Fig. 15.20. Originally a triangular pack (hence the name) the Tetra Pak is now produced in several shapes. This picture shows the Tetra Brik. (Courtesy Tetra Pak.)

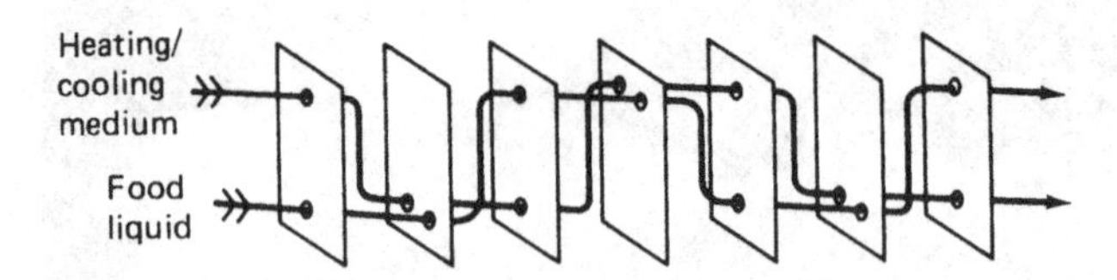

Fig. 15.21. Diagram of a plate pasteuriser. (Note: Horizontal bolts to hold assembly together and the plastic gaskets between the plates to prevent leakage are not shown.)

carefully otherwise the technological properties of the egg are affected. *Salmonella* are more heat-resistant in yolk than in whole egg due to the lower pH and higher total solids content in the yolk. For this reason the minimum pasteurisation requirements differ with the type of egg product. High-temperature storage of egg-white solids effectively destroys all *Salmonella* and a *Salmonella*-free product is therefore certain.

In spite of its general usefulness pasteurisation can have disadvantages for instance in the production of cheese. The coagulation and ripening of cheese requires a wide spectrum of microorganisms which are normally

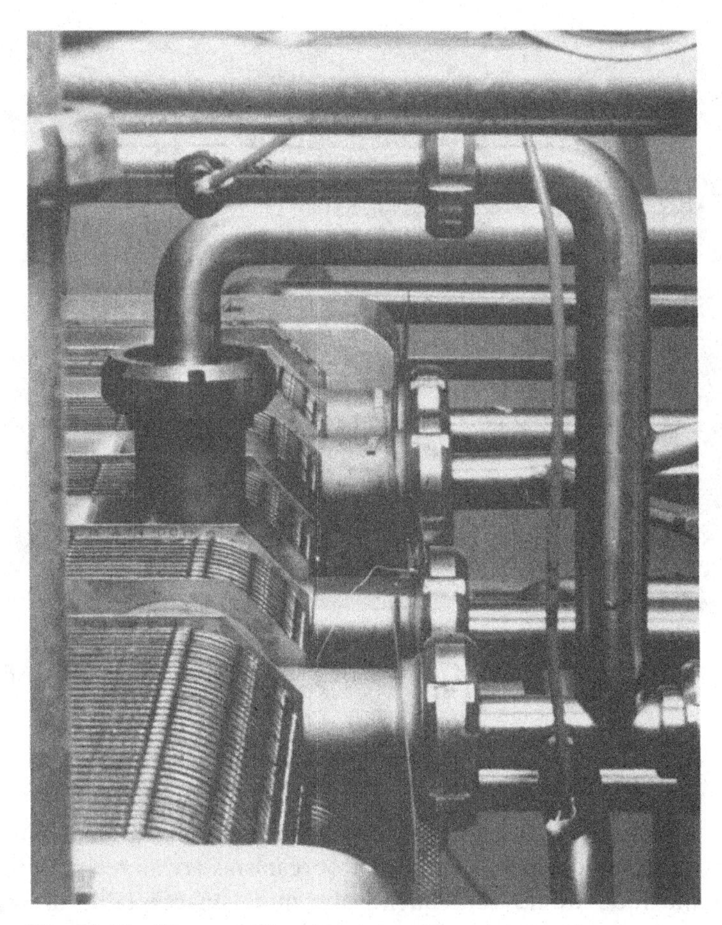

Fig. 15.22. Close-up of a plate pasteuriser showing the plates. (Courtesy Tetra Pak.)

contained in the milk. Pasteurisation destroys them and their artificial replacement requires a laboratory. Furthermore contamination after pasteurisation can spoil the cheese since the original lactic acid bacteria have been destroyed and the new mircoorganisms can thrive without competition. Pasteurisation can also encourage laxness on the part of the producer who may feel that cleanliness is no longer important. Finally during the heat treatment some vitamins are destroyed (about 30% of vitamin C) and some calcium may be lost through precipitation as an insoluble salt.

Fig. 15.23. Pasteurisation plant in a modern dairy. The plate pasteuriser is on the extreme right of the picture. (Courtesy APV Co. Ltd.)

Canning

When food is sterilised, both microorganisms and their spores are destroyed. Unfortunately some microorganisms are so resistant to heat that the food would subsequently be quite unacceptable as regards flavour, texture or nutritional properties. Therefore on canning, food is made 'commercially sterile'. Here pathogenic microorganisms are reduced to insignificant levels (typically by a ratio of 1 in 10^{12}) and a limited number of non-pathogenic microorganisms are left. These surviving organisms will not grow under the conditions of storage.

Originally foodstuffs were placed into containers and boiled in water to process them. Already in the last century was it realised that foods could be treated for shorter times if higher temperatures were used. The boiling point of water was first increased by adding salts e.g. calcium chloride. This reduced the cooking time of canned meats from six hours to about half an hour and so allowed production to be increased from 2000 to 20 000 cans per day. However, heat transfer in a solution is slower than when steam is used in a retort. Hence processing times must be longer and there is therefore more damage to the food. Furthermore, since the pressure in the cans builds up the cans can explode (Table 15.5).

Table 15.5 *Boiling points of aqueous solutions of calcium chloride at standard pressure*

Temperature (°C)	Calcium chloride (%)	Temperature (°C)	Calcium chloride (%)
100	0	110.0	29.3
101.7	8.5	115.6	36.3
104.4	18.5	121.1	42.0
107.2	24.5	125.0	45.8

simple control

A typical processing system for a pressure steam cook can be manually controlled as follows: the operator loads and locks the retort . . . turns on the steam . . . vents the retort . . . closes the drain . . . closes the overflow which serves as a vent . . . times the processing period after the retort reaches the set processing temperature . . . opens the compressed air valve . . . shuts off the steam . . . opens the water valves and manually controls the internal pressure of the retort during cooling . . . opens the drain valve . . . unloads the retort.

The control system automatically controls the processing temperature only and records it.

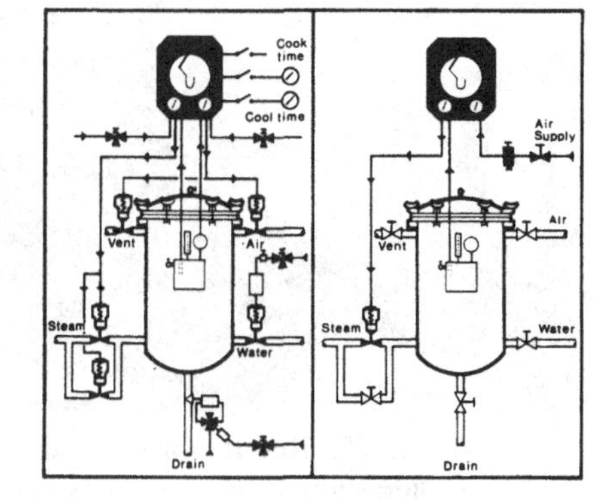

automatic control

The processing system can be fully automated, in which case the operator has only to carry out the following duties: set the cook temperature and time initially for the whole production batch . . . load and locks the retort . . . depresses the start button.

Automation does everything else until the time comes to unload the retort.
This system is advisable when glass or flexible containers are being processed in hot water with over pressure, since the operational sequence is critical.

Fig. 15.24. The retort and its operation. (Courtesy Mather and Platt (Engineering) Limited.)

Using pressure vessels the temperature can be further increased. The greater the pressure, the higher is the processing temperature. It is of course very important that both the steam used for sterilisation and the headspace in the can are free from air. Steam at 10 p.s.i. has a temperature of 115°C. As soon as it is contaminated with air, the temperature at the same pressure falls. When it is adulterated with 10% air the temperature will be only 112°C, at 20% air 108°C and at 40% air 100°C.

Figure 15.24 shows a diagram of a manually operated retort (left) and one with automatic control (right). A retort is similar to a domestic

Fig. 15.25. Retorts for batch sterilisation using steam from an outside source. (Courtesy Mather and Platt (Engineering) Limited.)

pressure cooker where the lid is screwed down allowing the interior to be heated under pressure above the boiling point of water. Figure 15.25 shows a series of industrial retorts. High-capacity continuous retorts are also used. For some foods e.g. rice pudding, it is necessary to rotate the cans during processing to increase the rate of heating and to avoid the contents from clumping together. Therefore rotary retorts are available. A small laboratory type is shown in Fig. 15.26.

Figure 15.27 shows a simplified canning line for vegetables. The clean vegetable is first sorted manually on the picking tale and blanched. It is then washed in the rod washer. This is essentially a trough of water with a rotating cage of metal rods. Often the washing precedes the blanching operation. The vegetable is now filled into cans and, if necessary, brine (or sugar syrup in the case of fruit) may be added. In the seamer the lids are fastened under vacuum to the cans which are placed into wire baskets or crates. These are hoisted into the retort. After commercial sterilisation the retorts are filled with cold water to reduce the temperature of the cans to a suitable level. The cans are finally dried, labelled, packed and dispatched. They should be stored at a relative humidity not exceeding 55–65% to avoid corrosion.

Fig. 15.26. Open rotary laboratory retort showing rotating basket containing cans.

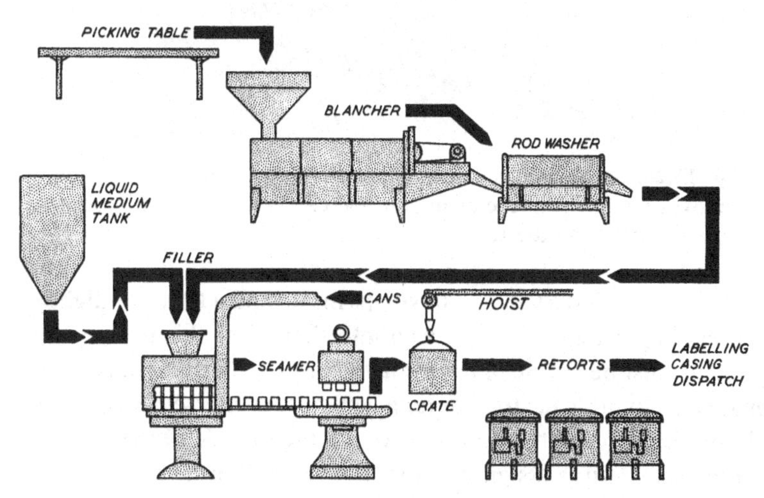

Fig. 15.27. A simplified canning line for vegetables (see text).

Fig. 15.28. A vacuum seamer. When the door is shut the machine removes the air from the chamber, and closes the lid on the can while it rotates on the turntable.

As the cans are cooled a vacuum develops inside and if they are slightly damaged, cooling water may be drawn into them. For this reason, the cooling water must be sterile and is usually heavily chlorinated. So if an occasional can should leak, the contents will not be infected. Cooling is stopped just above room temperature so that the outside of the can will dry and not rust. Rust is a sign of corrosion and will eventually lead to leakage and deterioration.

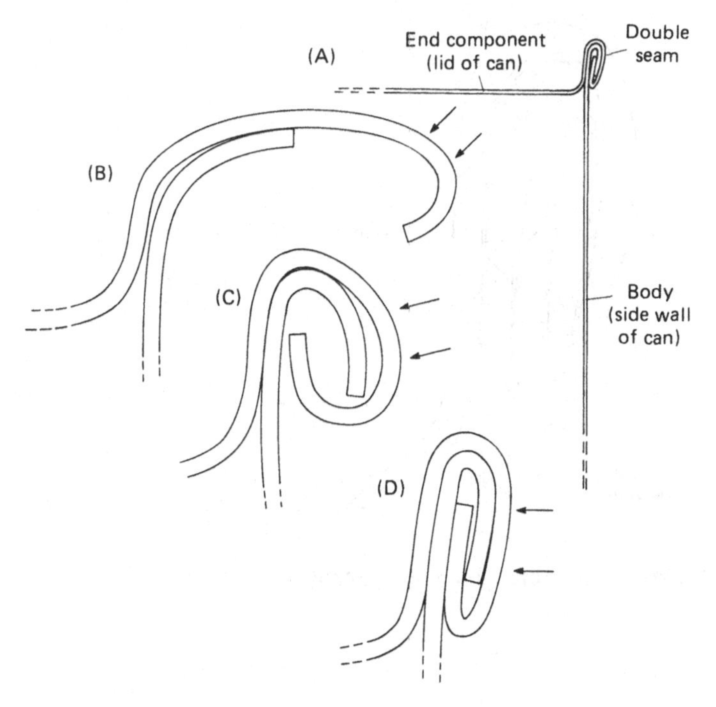

Fig. 15.29. The seaming operation.

One end of the can, the bottom, is already in place when the can is supplied by the manufacturer. This is referred to as the 'body'. The lid called the 'end-component' must be joined after filling the can and it is the canner's responsibility that the seam is a good one. Figure 15.28 shows a small vacuum-sealing machine. The chamber is open and the filled can can be seen on the rotating base plate. Above the can is the sealing chuck and rolls which fold over the lips of the lid and can and seal them firmly together (Fig. 15.29).

Figure 15.30 shows the resultant double seam. There should be adequate overlap and the seam should be tight (the can manufacturer will advise on this) and show no scuffed surface from the seaming rolls. If the inside of the can is lacquered to avoid corrosion by an acid content, there should be the least possible metal exposure at the seam. Different types of can are shown in Fig 15.31.

Clostridium botulinum will not grow and produce toxin at pH values below 4.5. Most fruits and berries have pH values below this and processing at 100°C is adequate for their preservation with safety. Therefore bottling in the home by placing these foods into glass jars fitted with hermetic seals (e.g. rubber rings) and boiling the jars in water is

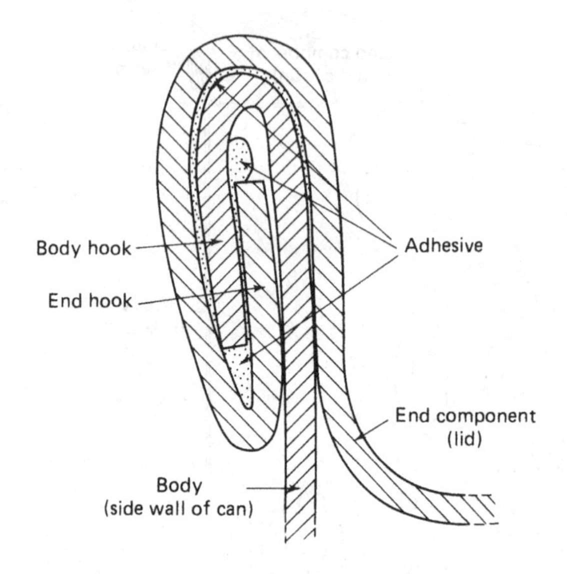

Fig. 15.30. Section through a double seam.

Fig. 15.31. Different types of can. Back (left to right): two-piece ring pull, three-piece ring pull showing long seam, and standard three-piece ('open top'). Front (left to right): flat key-twist, flat ring pull.

Table 15.6 *Processing temperatures for low- and high-acid food*

Type	Examples	Processing temperature
Low-acid foods, pH 4.5–7	Meat, fish, milk vegetables, soups	115–125°C
High-acid foods, pH 2–4.5	Fruits, fruit juices	100°C

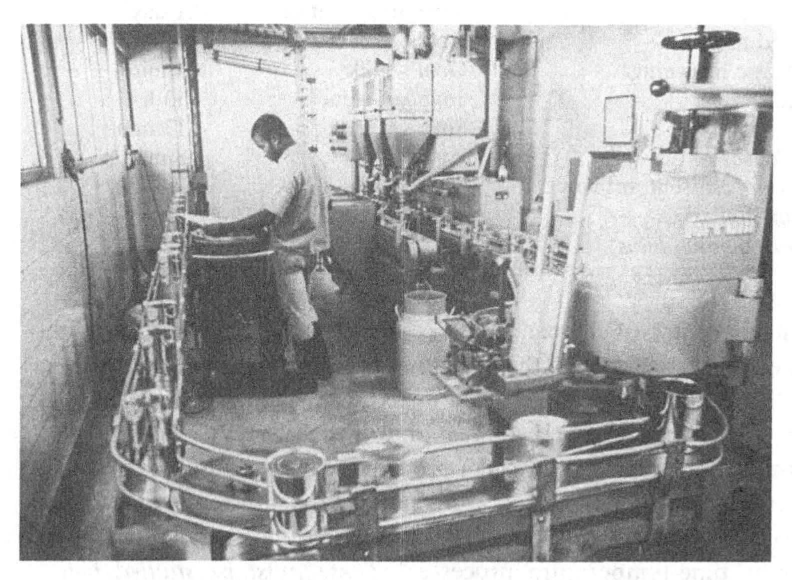

Fig. 15.32. Packaging line for a chocolate-flavoured drink. (Courtesy Nestlé.)

feasible. In contrast to these 'high acid foods', 'low acid foods' such as meat, fish, vegetable soups and dairy foods must be processed at higher temperatures, i.e. under pressure in a retort (Table 15.6).

The temperature at the centre of the can will approach the required processing temperature only slowly. Heat penetration on retorting can be determined by sealing temperature sensors (thermocouples) into the can. It will be found that metal containers allow faster heat transfer than polythene pouches or glass jars. Size and shape as well as the original temperature of the cans are important. Also, the contents of a rotating can heat up faster than those of a stationary one, particularly if the contents are liquid rather than solid. In the former heat transfer takes place by convection, in the latter by conduction.

With convection the coldest point of the can is slightly below the

Table 15.7 *Some* D-*values at 121 °C*

Material or reaction	Medium	D-value
Thiamin	Meat, vegetable	130–250 min
Thiamin	Vitamin preparation	1.5 days
Pantothenic acid	Vitamin preparation	4.5 days
Folic acid	Vitamin preparation	2 days
Vitamin A	Vitamin preparation	12 days
Vitamin B_{12}	Vitamin preparation	2 days
Vitamin C	Vitamin preparation	1 day
Meillard reaction	Apple juice	4.5 h
Enzymatic browning	Goat's milk	1 min
Lysine	Soybean meal	13 h
Trypsin inhibitor	Soybean milk	13 min
Peroxidase	Pea	3 min
Staphylococcus enterotoxin B	Milk	9.5 min
C. botulinum E.	Growth medium	4 min at 60°C
B. stearothermophilus	—	5 min

geometric centre of the can. With conduction the coldest point *is* the geometric centre although the insulating head space in the can will affect this. If the contents of the can are mixed, e.g. chunks of meat in thick gravy, both convection and conduction will occur simultaneously.

Apart from reaching the proper processing temperature it is of course important that this temperature is maintained for sufficiently long. There are well established procedures for calculating the sterilising effects of different time-temperature processes. *These must be studied before attempting a canning process*.

An indication of the sensitivity of different food components and microorganisms to thermal destruction is given by the D-value or the 'decimal reduction time'. This is the time taken at a given constant temperature for a reactant (e.g. a microorganism, an enzyme, a toxin or a vitamin) to decrease by 90% i.e. by one log value). The published D-values are only a guide because they are very variable and depend on various factors such as pH, oxidation-reduction potential, type of food, metal catalysts and on the actual temperature. Table 15.7 gives some D-values at 121°C. It is apparent that some microorganisms are destroyed quite quickly, while some vitamins, for instance, can be quite stable.

Irradiation

For about 30 years gamma radiation from a Cobalt-60 or Caesium-137 source has been used significantly to sterilise medical supplies such as

Table 15.8 *Recommended dose ranges for food irradiation*

Process	Range (kGy)
Inhibition of sprouting	0.05–0.15
Delaying ripening of various fruits	0.2–0.5
Insect disinfestation	0.2–1.0
Elimination of various parasites	0.03–6.0
Shelf life extension by reduction of microbial load	0.5–5.0
Elimination of non-sporing pathogens	3.0–10.0
Dried herbs and spices	10.0
Bacterial sterilisation	up to 50.00

instruments, linen and sutures. Cobalt-60 has a half-life of 5.3 years and Caesium-137 one of 30.1 years. Therefore the latter would have an economic advantage. In 1970 an international project of irradiation of food was instituted to establish safety. Since then irradiation has been used in a number of countries.

The unit of radiation is the Gray (Gy) which is defined as the absorption of 100 ergs of energy by one gram of material. Radiation doses for food are usually given as kilograys (kGy) where 1 kGy = 1000 Gy.

Table 15.8 gives the recommended dose ranges for food irradiation.

There are several food-irradiation systems. Figure 15.33 shows an automatic pallet irradiator. Two standard pallets loaded with product to a height of 1.80 m are placed by a fork lift truck into a carrier. This moves around the cobalt-source, which is heavily shielded, in a single pass. Both high and low dosages can be applied to the food.

The object of the treatment is to inhibit sprouting, destroy insects and parasites, decrease microbial load and extend shelf life.

An important effect of radiation is the breakdown of water molecules to give chemically active ions e.g. hydrogen atoms, hydroxyl ions, superoxide and hydrogen peroxide. As a result living cells are killed. Water is therefore part of the process and it would appear that the irradiation dose required is inversely proportional to water content.

The variation in density within the food package also affects the treatment and therefore the values given in Table 15.8 refer to 'overall average doses'. The variation between the highest and lowest dose in a sample is usually 1.5 kGy. Therefore the variation at 10 kGy would be between 8 and 12 kGy.

At that level all microorganisms are killed *but not their spores* which have a low water content. Hence complete elimination of *Salmonella* requires only 2–8 kGy but that of *Clostridium botulinum*, a spore-former, 50 kGy. Irradiated foods which are capable of supporting the growth of

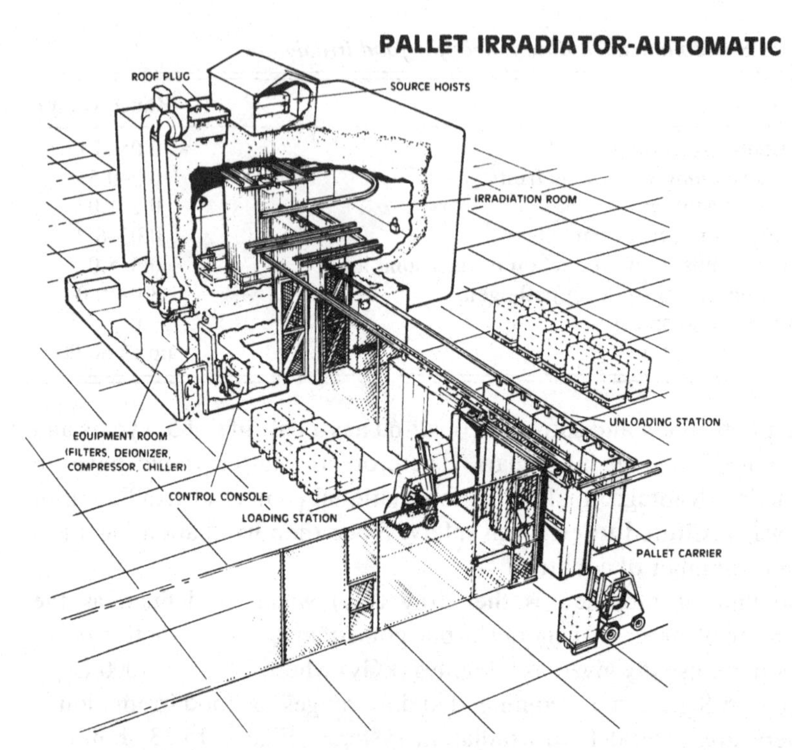

Fig. 15.33. Automatic Pallet Irradiator. (Courtesy AECL Irradiation Division.)

microorganisms must be kept at a sufficiently low temperature to prevent the new growth of pathogens the spores of which have survived e.g. *Cl. botulinium*.

There appear to be no significant nutritional problems below 10 kGy. There may be damage to vitamin E and in the presence of air to unsaturated fatty acids but no more than with other methods of preservation (e.g. freezing, pasteurisation, canning). It has been claimed that meat may develop a 'wet dog' odour and milk a burnt taste. There is also no significant induced radioactivity or chemical change in the food. However, at the doses required for sterilisation there are significant changes in the product and therefore at the moment irradiation does not appear to be an alternative to canning.

Radiation for chicken, fish, rice and legumes is now used in Bangladesh, Brazil and Chile. Irradiated mangoes, pawpaw and strawberries have been sold in South Africa and apparently plants for the irradiation of prawns are being built in Paraguay and Thailand. There are also several plants in Holland and France.

16

Food additives

Food additives are substances which are added to the basic foodstuff before it is consumed. It has been suggested that there are as many as 10 000 additives in use. If that is an exaggeration, there are certainly very many. In developed countries there are organisations which test, control and legislate in an attempt to make certain that an additive is no hazard to health. Since testing is costly, time-consuming and often does not give clear answers (one cannot use human beings for dangerous tests!), there are differences in the various lists of permitted substances. There are permitted lists applicable to the UK, US, EEC, FAO/WHO and many others. Developing countries often follow one or other of these lists.

Food additives are generally useful, sometimes unnecessary but must be objected to if they help upgrade for sale poor or even spoilt and unhealthy food.

Food additives are not as common or important in developing countries as they are in developed ones. However great vigilance must be exercised because unscrupulous dealers may dump an additive in a developing country because it has been declared unsafe in a developed one and surplus stocks are cheap.

Additives are difficult to classify because some have more than one function; some acids both preserve food and add flavour. Bread improvers may bleach and increase loaf volume. Sugar or salt may flavour the food or at higher concentrations act as preservative or a flavour. So any classification of food additives is a little arbitrary.

Preservatives

These are substances which prevent food spoilage, which is usually due to microbial or enzymic activity. Antioxidants and chelating agents (see below) can also be included under preservatives.

Benzoic acid and its sodium salt, sulphur dioxide, sulphite and bisulphite are often added to acid fruit and their juices, to pickles and soft drinks. Benzoic acid which is also sometimes added to margarine, has been known to produce allergies in some people. Sulphur dioxide is very pungent and may also affect the colour (mainly carotenoids) of the food and break down the vitamin folic acid.

Lactic acid, sodium diacetate, sorbic and propionic acids as well as their salts are used as mould inhibitors in baked products. Sorbic acid is also used in cheeses, syrups and pie fillings.

Antibiotics (penicillin, streptomycin, nisin, nystatin) have been used to counteract bacteria but do not act on fungi. Their use is often prohibited because they kill the human intestinal flora (important e.g. in vitamin metabolism) and cause loss of resistance to pathogenic bacteria.

Ten to twenty per cent carbon dioxide inhibits both microorganisms and the development of oxidative rancidity in chilled meat. Since the cold rooms must be built air tight they are expensive to construct.

Salt (NaCl) is a very ancient preservative. It is used mainly with meat and fish. At a level of 10% it is bacteriostatic (i.e. it stops bacterial growth) and withdraws liquid from the food. It must, however, be remembered that some bacteria and fungi are halophyllic (salt-loving). These are tolerant to salt or even grow best at high salt concentrations. Often nitrate and nitrite are also added with the salt. Nitrate is converted into the nitrite by microbial agency. The nitrite gives an attractive red colour to meat and is also very effective against *Clostridium botulinum*. Unfortunately it also gives rise to some nitrosamine. This has been shown to cause cancer in rats. Whether it causes cancer in man is at present not clear.

Smoking is also a very ancient method of preservation mainly for meat and fish. There are basically two methods. The first is cold-smoking. Here the temperature does not exceed 30°–40°C and protein is not denatured. This method is not really feasible in the tropics. The second method is hot-smoking which combines three effects; cooking, dehydration, and the preservative effect of the smoke. Therefore on smoking there is a weight loss and a distinctive change in taste and colour. Often these changes are more important than the preservative effect.

Traditional smoking arrangements are usually not suitable for large quantities of marketable produce. A brief survey is given in Table 16.1. Although the smoking ovens are usually cheap to build they tend to have a low capacity. Fuel consumption is high, they require constant attention, are difficult to control and are often affected by wind and rain.

With modern methods smoke density, air velocity and humidity are

Table 16.1 *Traditional smoking ovens*

Type	Materials required
Open pit	Pit with wooden platform above
Drum oven	Half or whole 200-l oil drum with wire mesh above
Stone oven	Stones around fire with wire mesh above
Smoke baskets	Wire basket or perforated clay pot over fire
Smoke house	Roofed building with purpose-built platform and firing arrangement

controlled. Sawdust is heated by electricity or superheated steam. The type of smoke depends on the type of wood used. Hard woods contain rather more lignin and tend to produce bland flavours and little colour in the product. Soft woods tend to contain more cellulose and give rise to more acid flavours and stronger colours. So far 130 compounds have been identified in wood smoke and some of these at very high doses have caused cancer in rats. A significant reduction of these potentially harmful compounds can be achieved by filtering the smoke through cotton or steel wool. Recently 'liquid smokes' have been used. These are solutions containing some of the identified smoke constituents. So far the flavours produced by these methods are not quite satisfactory but they have the advantage that known possible carcinogens can be omitted from the mixture. Salt and spices may assist smoke preservation but the product must still be kept dry after treatment or moulds will develop.

Antioxidants are compounds which inhibit oxidative rancidity in fats. Among the antioxidants used in the food industry are vitamins C and E which occur naturally. Among the synthetic antioxidants are butylated hydroxyanisole (BHA) and butylated hydroxytoluene (BHT).

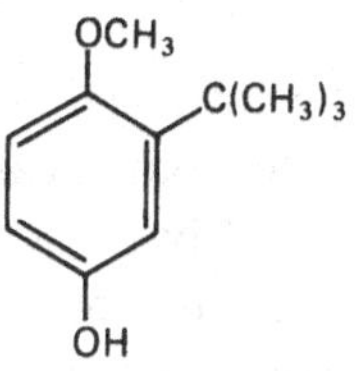

BHA

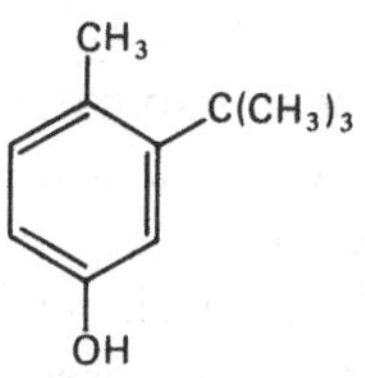

BHT

The words 'chelating agent' come from the Greek word chele meaning a crab's claw. The alternative term for these chemicals 'sequestering agents' comes from the Latin sequestrare which means to place in safe keeping. Trace elements such as iron and copper often speed up the deterioration of food. The destruction of vitamins C, E, folic acid and thiamin, the development of pink colours in canned pears or the grey-green in canned maize, clouding in soft drinks, chill haze in beer, and the formation of struvite (ammonium magnesium phosphate) crystals in canned fish are all accelerated by these metals. Chelating agents effectively bind them without interfering with human trace-mineral metabolism. Important chelating agents are ethylene diamine tetra-acetic acid (EDTA), tartaric and citric acids and their salts. EDTA is used clinically to treat metal poisoning. A very simply chelating agent is glycine which can form a complex with e.g. copper.

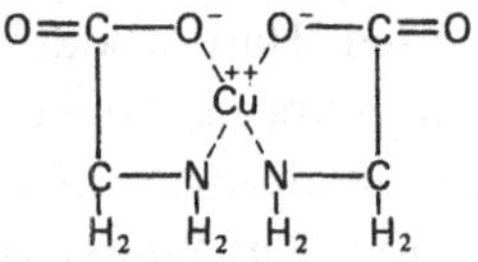

```
O=C—O⁻  O⁻—C=O
  |   \++/    |
  |    Cu     |
  |   /  \    |
  C—N      N—C
  H2  H2   H2  H2
```

Colouring agents

The colour of food is most important to the consumer. When a food has lost its colour in processing it is reasonable for the manufacturer to use a colouring agent to make it more attractive. It is not acceptable if the colour is used to conceal damage or inferiority and make the food look better than it is.

When coal tar dyes were discovered in the last century, the food manufacturer thought that his colour problems were solved because very small amounts of dye would give a deep, consistent and pure colour. Unfortunately several of these have been shown to be carcinogenic and the list of permitted colours has now been greatly reduced. Very extensive testing is now required by many countries before a colour is permitted.

Colours can be divided into natural, synthetic, and artificial. The first are obtained from natural sources. Examples are chlorophyll, carotene, tumeric, anatto, cochineal and caramel. The second have been synthesised in the laboratory and include carotene, riboflavin and curcumin.

The great majority of dyes used are artificial. They tend to be more stable and uniform as well as cheaper than the natural products. The testing of these colours is very laborious and expensive requiring metabolic studies with several species of animal, including man. The various authorities are not agreed as to how exactly to test the food colours with the result that most permitted lists differ. FAO/WHO lists 31 dyestuffs, while the UK list also contains 31 but the two lists differ considerably.

In this connection one should also mention chemicals which are used to remove colour. The most common is benzyl peroxide used to bleach flour. Other flour additives e.g. chlorine dioxide and chlorine have both a bleaching and 'improving' effect, i.e. they improve loaf volume and texture. Bleaching agents may also be used in cheese.

Flavouring agents and enhancers

There are an extremely large number of flavouring substances used in the food industry. Sometimes a particular flavouring agent is confined to one country or continent but more often it is very widespread. Because there are so many flavours few of them have been tested and some natural spices or flavours may easily contain a mixture of 50 different chemical components. About 500 flavours are referred to as natural. These include lemon and orange oils which are extracted from the rind, as well as plant leaves, fruits, roots, bulbs and tubers. Thirteen of these may possibly be toxic and should be banned.

The great majority of flavours are artificial and synthetic. Since it would be impossible to test them all, one must exercise one's judgement. About 250 of these flavours have chemical formulae which may suggest that the substances are not too healthy if consumed in excessive quantity. A great deal of testing is still required.

There are also flavour enhancers. The most common and oldest is sodium chloride. More recent ones are monosodium glutamate (GMS), ethyl maltol and disodium inosinate and guanylate.

Micro-encapsulation

Tiny particles of solid, liquid or gas can be surrounded by a shell or capsule. In this way e.g. liquids can be converted into free-flowing powders. When the particle size of the capsules is less than 500 μm the particles are referred to as microcapsules and the process is called micro-encapsulation (Fig. 16.1). The material used to encase the core material is known as the wall or capsule. Choice of the wall material depends on the type of material to be encased and the release properties required. The most widely used wall materials are vegetable gums (e.g. gum acacia, guar gum), starches or modified starches, dextrins, proteins (e.g. gelatin,

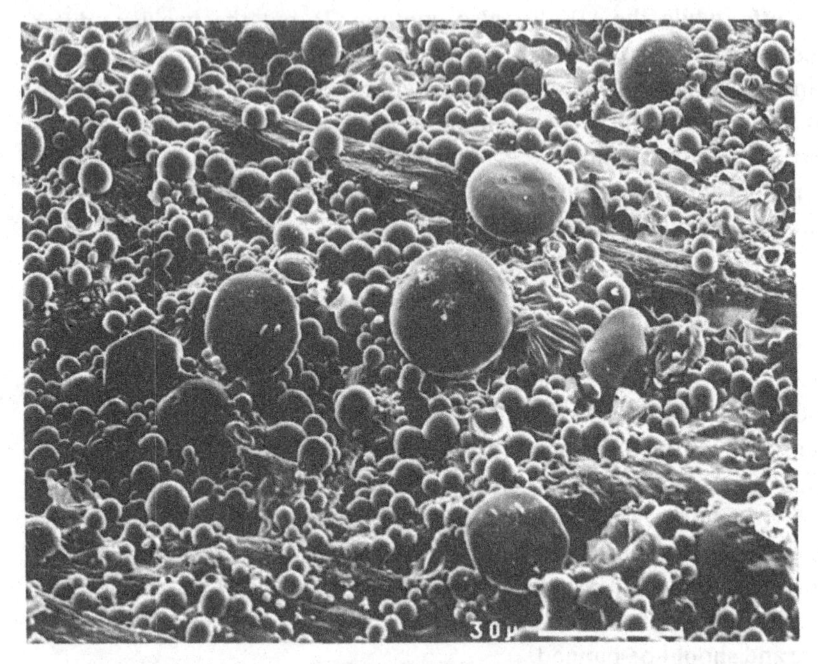

Fig. 16.1. Scanning electron micrograph of microcapsules attached to paper. Damage to the capsules on handling the paper can release odour (see text). Bar = 30 μm.

hydrolysed gelatin, soya protein, caseinates), cellulose esters and ethers and sugars (e.g. sucrose, glucose).

Often a combination of these is used. Other materials such as emulsifiers, surfactants, chelating agents or plasticisers may also be included. When dealing with food additives, it is a necessary requirement that the wall material should be edible and non-toxic. Also, the capsule should not react with the encased material or other food components present.

Several methods are used to manufacture microcapsules but the most common is spray-drying.

At first the wall material is dissolved or dispersed in a solvent, usually water. The core material, an essential oil or oleoresin is emulsified by mechanical means into the wall material. The emulsion is then spray-dried and this causes the dispersion to break into small droplets consisting of core particles, individually surrounded by a film of the wall material. The solvent is rapidly removed from the shells resulting in the formation of dry capsule walls around the core material.

Thus the core material is protected from loss of volatiles, degradation and oxidation and the microcapsules are readily incorporated into the food or on to the wrapping. Shelf life is improved and the flavour is quickly released on cooking or eating.

Texturing agents

Particularly in hot and dry climates, humectants are often used to prevent foods drying out. Glycerol and sorbitol have been used with shredded coconut and soft cakes. The opposite function – to keep foods (e.g. salt) dry and free-flowing is performed by anticaking agents. These include calcium phosphate, magnesium oxide and salts of silicic acid. Some calcium salts (e.g. the hydroxide, lactate and phosphate) also act as firming agents by keeping vegetable tissues turgid. For instance canned tomatoes are stabilised in this way. Gums of which there are many are used as texturisers in milk drinks and ice-cream as well as in beer as anti-foaming agents.

Surfactants like gums must be carefully chosen. Some prevent seepage of water from fats and cheeses and others prevent fat-bloom in chocolate. Some reduce the stickiness of peanut butter or caramel while others stabilise foam such as mousse or meringue. Some gases such as nitrogen, carbon dioxide and nitrous oxide have been used as whipping agents in aerosol bottles.

Processing aids

Perhaps the most important of these are enzymes and the use of these is very ancient (see also p. 22). Barley malt has been usd in beer manufacture for centuries and the first enzyme isolated, diastase, was obtained from it. Today barley malt is still an important source of amylase. A well known enzyme of animal origin is rennin used for curdling milk in cheese production and obtained from the calf's stomach. Papain isolated from the unripe papaya fruit, bromelin from pineapple and ficin from figs have a proteolytic effect and are used to soften meat or completely hydrolyse meat or fish. Modern enzymes, many of microbial origin, are used for such different processes as starch degradation, fruit juice clarification, desugaring of eggs, oxygen removal from foods liable to oxidation, cleaning of shellfish, degreasing bones, solubilising tea solids and removal of diacetyl from beer (Table 16.2).

Other processing aids are ethylene to speed up ripening of bananas and malic anhydride to prevent sprouting in some tubers. Oxidising agents (bromate, iodate) in bread manufacture have already been mentioned. They make the dough less sticky and better to handle as well as increasing

Table 16.2 *Use of selected enzymes in food production*

Enzyme	Source	Application
α-amylase	*Bacillus* spp. *Aspergillus* spp. *Rhizopus* spp. Pancreatic tissue	Breaks down starch to simple sugars. Used in production of syrups. Controls sugar ratios. Predigests baby foods. Accelerates fermentation.
β-amylase	*Bacillus* spp	Similar to α-amylase but produces maltose.
Amyloglucosidase	See Glucoamylase	
Catalase	*Micrococcus lysodeicticus*	Breaks down H_2O_2, prevents oxidation
Cellulase	*Aspergillus* spp	Breaks down cellulose. Improves yield of juices and flavour extracts.
Ficin	Fig	See papain.
α-galactosidase	Almonds, yeast	Reduces flatulence factor.
β-glucanase	*Rhizopus* spp.	Breaks down β-glucans, reduces viscosity to aid filtration
Glucoamylase	*Aspergillus* spp. *Rhizopus* spp. *Trichoderma*	Degrades carbohydrates, increases fermentability.
Glucose isomerase	*Bacillus coagulans* *Streptomyces* spp.	Converts glucose to fructose, increases sweetness.
Glucose oxidase	*Aspergillus* spp.	Removes oxygen, reduces non-enzymatic browning.
Invertase	*Saccharomyces* spp. *Klyveromyces* spp.	Converts sucrose to glucose and fructose, increases sweetness.
Lactase	*Klyveromyces* spp.	Breaks down lactose. Used in dairy industry.
Lipase	*Aspergillus* spp. *Mucor* spp. *Rhizopus* spp.	Splits fats, helps remove or reduce fats, improves flavour in milk and cheese.
Lipoxidase	Soya	Bleaches wheat flour pigments and improves flavour in bread making.
Lipoxygenase	Various plant seeds	Bread improver.
Papain	Pawpaw	Softens dough, tenderises meat, reduces viscosity of fish protein dispersions and reduces gelling. Reduces chill haze in beer.
Pectinase	*Aspergillus* spp.	Breaks down pectin, reduces viscosity thus increasing filtration rates and fruit juice yield. Clarifies fruit juice.
Pepsin	Stomach	See papain.
Rennin	Calf stomach	Precipitates milk protein in cheese manufacture.

the bread volume. Clarifying agents such as pyrolidone and tannin in brewing and gelatin and albumin to precipitate iron and copper are also in frequent use.

Nutritional supplements

Food enrichment is defined as the replacement of nutrients removed during processing. Examples are the replacement of vitamin C in potato powder (in which it has been largely destroyed on dehydration) or the addition of thiamin to white flour (from which the vitamin has been removed in milling). *Food fortification* means that additional nutrients have been added to the food to make it more nutritious than it was before. Examples are the addition of iodine to table salt to prevent goitre or the addition of iron to wheat flour to help prevent iron-deficiency anaemia.

It is also possible to raise the quality of a synthetic product to that of the natural product. (An example is the addition of vitamins to margarine to make it nutritionally equivalent to butter.)

In developed countries regular food surveys are undertaken. The food is varied and the authorities are aware of possible deficiencies. For that reason the basic foodstuff, often wheat bread, is enriched with thiamin, riboflavin, niacin, iron and calcium. Water may be fluorinated (to help prevent tooth decay) and salt may be iodised.

In developing countries supplementation is often not as well based scientifically. Detailed dietary knowledge is not available and there may be lack of money. Good progress has, however, been made in rice-eating areas (Asia) to defeat beriberi and keratomalacia (deficiency of vitamin B_1 and A respectively), by rice supplementation. Cereals are the world's major food staple and are often used as a base for enrichment. Protein enrichment has included non-fat dried milk, fish and soya protein concentrates as well as microbial protein (see composite flours p. 119).

17

Food, hygiene and health

The relation between food and health can be divided into two aspects. The first deals with materials one eats or drinks, and should not. These are food and water contaminants and would include microorganisms, carcinogens and organic and inorganic poisons.

The second concerns nutrients which may be deficient. That could be shortage of protein or energy, or vitamin, mineral or fibre deficiency. In this chapter some of the aspects will be considered which are of special importance in tropical countries.

Water and food contaminants

The contamination of water depends often on its source. In developed countries clean and wholesome water is usually supplied. In developing countries that is not always so and water treatment may have to be carried out by the individual user. There are several sources of water. Rainwater is highly aerated and may contain traces of gases from industrial sources (ammonia, sulphur dioxide). It is usually low in salt but may be contaminated by the surfaces on to which it falls. For example bird droppings on roofs may contain *Salmonella* and *Clostridium*.

Spring water has percolated through the ground and escaped at a lower point. Depending on the soil it may contain salts. Since it has been filtered through the soil dangerous organic matter is usually removed and springs can therefore provide the best kind of drinking water.

Deep-well water is also often good provided that the walls have been well built and there has been no contamination from above. Water from shallow wells is often highly contaminated. River water may also be highly contaminated with sewage. However, microorganisms and plants may remove poisons and if pollution is limited the water may possibly be fit for use.

Sources of bacterial contamination of water are almost always through sewage, i.e. faecal contamination. Recent fieldwork carried out in Nigeria and Zimbabwe has shown that faecal contamination of water by animals can be identified by detection and counts of *Rhodococcus coprophilus* and that of human contamination by estimating sorbitol-fermenting *Bifidobacteria*.

Water may also become contaminated through excessive mineral content. This is not usually dangerous because the water soon becomes too unpleasant to drink. In some mining or industrial districts water may also be contaminated with effluents containing copper, lead, zinc, arsenic or mercury. However by far the most important contaminants are life forms.

There are four categories of disease which may occur:

(1) Infections spread through water supplies, i.e. water-borne diseases such as typhoid, cholera and infectious hepatitis.
(2) Diseases due to lack of clean water for personal hygiene. These are referred to as water-washed infections. They include scabbies, trachoma and bacillary dysentery.
(3) Water-based diseases. These are infections transmitted through aquatic invertebrate animals either penetrating the skin such as in schistosomiasis or through ingestion, e.g. of Guinea Worm.
(4) Infections spread by water-related insects either biting near water (sleeping sickness) or breeding in water (yellow fever, malaria).

Water taken from a source contaminated by life forms requires sterilisation. The most effective way is by boiling. Porcelain filters should never be used. They soon become clogged unless the porcelain fitments are scrubbed and boiled every few days. Even then the filters may be reassembled the wrong way and water may pass unfiltered. Boiled water, cooled in a clean receptacle is usually suitable for domestic use. Chlorination is good for community supply, but not easily done in the home because it requires expert knowledge.

The microbial contamination of food has already been discussed in Chapter 6 and it has been pointed out that food infections and food poisoning caused by bacteria are usually due to low standards of hygiene. Similarly mycotoxins produced by field or storage fungi can be a serious hazard to health particularly in the wet tropics (p. 95). Higher plants also contain toxins but often at extremely low levels. These levels are so small that one need not worry unduly and much more research requires to be done. An important exception is cassava (p. 142).

Possible poisons introduced through food additives have been considered in Chapter 16. These are not as important in developing countries as they are in developed ones because there are not so many additives about. Nevertheless the possible dumping of substances prohibited in the West must be carefully monitored.

Another problem in some developing countries is the use of prohibited pesticides for purposes for which they were not intended. A few years ago some villagers near Lake Volta (Ghana) found that HCH (hexachlorocyclohexane), which is a pesticide prohibited in many countries, was useful in stunning fish when sprinkled on the water. It caused reduction in the fish population and in man convulsions and possible birth defects and cancer. To use a hazardous pesticide for its true purpose is very undesirable. To use it for a task for which it was not designed is the height of folly.

Nutritional deficiencies

Protein-energy malnutrition (PEM)

This term describes a range of diseases. At one end is marasmus. This is due to a grossly inadequate intake of food. Those suffering from it show the typical features of starvation. Tissue fat and protein are metabolised to provide energy and as a result there is an emaciated appearance with a wrinkled or folded skin. Marasmus may begin with an abrupt early weaning at about five months after birth, perhaps as a result of copying Western practice. The infant is introduced to commercial milk replacements or weaning foods. These are not harmful in themselves but because of poverty or lack of education they may be over-diluted or made up with contaminated water. The infant may be subjected to repeated infections leading to diarrhoea, vomiting, dehydration and circulatory failure. Death may also occur due to low body temperature particularly at night. This is due to low energy reserves and poor tissue insulation through lack of subcutaneous fat. Treatment is the restoration of a well balanced and adequate diet.

At the other end of the range is kwashiorkor. This is due to low protein intake although total food intake may be adequate. It often follows late weaning between 1 and 2 years and the introduction of a low-protein diet, e.g. based on root crops (cassava, yam, cocoyam). There is muscle wastage obscured by swelling (oedema) particularly in the legs. This is due to excess body water in the extracellular fluid space. The skin may change colour, becoming paler or darker. The hair may become reddish in colour. The infant becomes very weepy. Death may occur through secondary infections such as gastroenteritis or measles. Treatment is a low restoration of a full diet avoiding overloading of the digestive tract. Milk is invaluable. Rest is also important and attendant infectious diseases must be treated.

Marasmus and kwashiorkor are the extremes of a range of conditions. Between them are intermediate states due to variable deficiencies of protein, energy, vitamins, essential fatty acids and minerals with possible effects of infectious diseases, parasites and possibly mycotoxicoses.

Vitamin deficiencies

Due to increases in nutritional knowledge vitamin deficiencies have decreased the world over. However the following are still found.

Beriberi This disease is due to a deficiency of vitamin B_1 (thiamin). It was formerly a major problem in south-east Asia. Thiamin is contained in the outer layers of cereals and these are removed in milling. Hence populations which subsist largely on polished rice are liable to the disease. Beriberi occurs in two main forms, wet and dry. In the former the body may be swollen with fluid (oedema). This is not present in the dry form. In both forms paralysis of the limbs may occur and death may follow through heart failure. The diagnosis is not always easy but if there are several people with these symptoms in a predominantly rice-eating area suspicion of the disease should be aroused. Infantile beriberi may be caused by thiamin deficiency in the mother's milk. To treat the disease, thiamin should be given by mouth or injected. Yeast, rice polishings, and eggs should be freely given.

Pellagra This is due to a deficiency of nicotinic acid (niacin) and may occur where maize is the staple food. It used to be quite common in southern Africa, but is rarer now. The disease is characterised by the 'three D's': Dermatitis, an itchy inflammation of the skin where it is exposed to the sun; Diarrhoea, of chronic but moderate severity; and Dementia, mental disorientation. The patient may be unwilling to co-operate. The tongue may be inflamed and bright red in colour. The disease is treated by oral administration of nicotinic acid. Since this itself is a powerful drug several small doses not exceeding 300 mg/day should be given. A mixed diet prevents the disease.

Riboflavin deficiency In a mild form this is fairly common in the poorer regions of the world. There are cracks and fissures at the angle of the mouth, soreness of the tongue, roughening of the skin of arms, shoulders and legs. There may be inflammation of scrotum or vulva. The eyes are very sensitive to light. The disease is treated with riboflavin or food rich in it.

Scurvey Minor degrees of the disease due to lack of vitamin C are quite common in the tropics. A patient with fully developed scurvey is very ill. His teeth are loose and the gums swollen and bleeding. There are small haemorrhages under the skin and larger ones into joints and muscles. He may be very anaemic, his heart weak and there may be oedema of the legs. Treatment consists in providing him with synthetic ascorbic acid (vitamin C) or fruit containing vitamin C.

Rickets This disease is caused by a deficiency of calcium. This can be due to a deficiency of calcium itself or to a deficiency of vitamin D which is involved in calcium metabolism. In some parts of West Africa and India there is a poor supply of calcium in the diet. Every effort must be made to obtain as much edible calcium as possible through chalk, bones of fish and other animals, leaves of high calcium content or their ash. The use of dairy products should be increased particularly for young and growing people. The normal supply of vitamin D is obtained either through the diet or from synthesis in the skin. Sunlight falling on the skin produces the vitamin. Rickets are most common in children and women where religious customs forbid them to walk abroad unless completely covered with clothing. The rickety child is fretful and has a large head with a failure of the bones of the skull to join. Pigeon chest, protruding abdomen, deformed pelvis and most noticeable of all bent legs are common.

The adult counterpart of rickets is called osteomalacia and may occur in women where there is undue loss of calcium through repeated pregnancies, lactation, lack of sunshine or vitamin D. As there is decrease of ionised calcium in the blood plasma muscular weakness or cramp is common. Withdrawal of calcium from the bones may cause sudden fracture.

Treatment for rickets and osteomalacia consists in providing adequate supplies of vitamin D and calcium.

Xerophthalmia–Keratomalacia There are large stores of vitamin A in the liver but if there is a shortage for over one year deficiency symptoms appear. The mildest is night blindness. This is difficult to diagnose but is usually detected in a community by an unusually high incidence of minor accidents at night.

The first clear sign of deficiency is xerophthalmia. The conjunctiva covering the white of the eye itself has a smoky grey colour. The conjunctiva shows greyish or white spots of thickened cells (Bitots spots). Eventually the cornea becomes dull and hazy. Finally the cornea liquifies, softens and ulcerates. This is keratomalacia which may lead to permanent blindness. In school children the skin may become roughened ('toadskin').

Deficiency of vitamin A is common in the tropics and the most common cause of blindness in the world today. Treatment consists in an adequate supply of vitamin A.

Mineral deficiencies

Of the possible mineral deficiencies only three are of sufficient general importance to be considered here. These are iron, calcium, and iodine deficiency.

Iron and calcium deficiency Iron deficiency is probably the most common nutritional deficiency disease in the world. However the large majority of cases of anaemia are so mild that the sufferer would not seek medical help. Iron is widely found in foodstuffs and is required only in small amounts (0.5–1 mg/day absorbed iron). Unfortunately only about 10% of the iron ingested is absorbed.

The total daily losses are also quite small, 0.1 mg/24 h in urine and a total of 0.5 mg from all sources, except during menstruation which produces a loss of about 28 mg per menstruation. However, the balance between intake and output is quite fine and a slight decrease in absorption or increase in excretion is sufficient to place the person in negative balance with a risk of anaemia.

Iron deficiency is probably caused not so much by deficient intake as by poor absorption or increased loss of iron. Fortunately the absorption of iron is increased when body tissue concentration decreases and because iron is little excreted, any deficiencies tend to be rather mild. The person is also buffered against severe anaemia by the presence of large stores of iron (ferritin) in the liver. Iron from damaged red blood cells is also reused.

If prolonged, mild deficiencies can become much more severe and even fatal. Because of the significant blood losses in menstruation and in childbirth women are in greater danger of developing anaemia than men. Because of its tendency to reduce menstrual blood loss, the introduction of oral contraception has decreased the incidence of anaemia in women.

Because of the role of iron in the synthesis of the oxygen carriers of respiration (e.g. haemoglobin), anaemia results in impaired delivery of oxygen to the sites of metabolism. The normal level of haemoglobin varies with sex or age but it is usually greater than 11–14 mg/100 ml of blood. In anaemia this decreases to less than 7–8 mg/100 ml blood and the red blood cells are small and pale because they are deficient in haemoglobin. The outward signs of this are general fatigue and lassitude. Because the onset is so gradual in mild anaemia, it may be hardly noticeable particularly in those involved in sedentary activities. During exercise, however, the increased demand for oxygen and the inability of the anaemic individual to meet it fully, appear as extreme fatigue and breathlessness.

Because of the similarity of the effects of deficiency of calcium and of vitamin D the reader is directed to the section on vitamin D deficiency.

Iodine deficiency Iodine deficiency is seen in those areas of the world where the iodine content of the soil and hence that of the vegetation is exceptionally low.

Iodine is required for the production of thyroid hormone in the thyroid gland. As the thyroid concentration of iodine decreases, the production of hormone decreases. In an attempt to increase the production of hormone, the thyroid gland enlarges. Thus the most obvious sign of iodine deficiency is a swollen neck caused by the grossly enlarged thyroid gland. The deficiency is called goitre.

If it is very severe and present from an early age, up to 5% of individuals may show signs of mental retardation or cretinism. This mental retardation may be associated with dwarfism (possibly because of an interaction between thyroid and growth hormone), and with oedema.

While the commonest cause of goitre is iodine deficiency, certain foods contain agents which bind iodine and make it unavailable to the body. These are called goitrogenes. They are contained in some varieties of cabbage, cassava and other vegetable foods. These foods can cause goitre if eaten in large quantities.

Goitre has been almost totally eradicated in the developed countries of the world by the addition of iodine usually as potassium iodide to table salt. Today goitre is found in some areas of East Africa, Thailand, Malaysia and Indonesia.

Other diseases connected with food

The diseases discussed so far are well understood and there is little scientific argument about them: e.g. vitamin C deficiency causes scurvey, iodine deficiency causes goitre, *Salmonella* cause food poisoning. However, there are some areas which are not well understood and scientists still argue about them. A problem in some countries is overweight (obesity). This is due to excess fat in the body. First it is difficult to measure excess fat and second it is not clear how harmful obesity is. Similarly coronary heart disease is a condition where the artery supplying the heart muscle itself becomes blocked by deposition of fatty materials. The result is heart failure and death. It has been claimed that this disease is prevalent in affluent industrialised societies and associated with a high intake of sugar and fat.

It has also been said that the high intake of fibre usual in developing countries protects against various intestinal disorders including cancer, which are becoming increasingly common in developed countries. Although there is evidence that fibre deficiency is harmful the evidence for it causing various diseases is not conclusive.

In any event research into these problems is mainly the task of developed countries and not so important for developing ones.

APPENDIX 1

Scientific names of some of the plants and animals mentioned in this book

Cereals

Adlay	See Job's tears
Barley	*Hordeum* spp.
Bullrush millet	*Pennisetum typhoides*
Finger millet	*Eleusine coracana*
Job's tears	*Coix lachrima-jobi*
Maize	*Zea mays*
Oats	*Avena sativa*
Rice	*Oryza sativa*
Rye	*Secale cereale*
Sorghum	*Sorghum* spp.
T'ef	*Eragrostis abyssinica*
Wheat	*Triticum vulgare*
Wild rice	*Zizania aquatica*

Legumes

Bambara groundnut	*Voandzeia subterranea*
Bean	*Phaseolus* spp.
Bean, black eye	see Cow pea
Bean, butter	*Phaseolus lunatus*
Bean, calabar	*Physostigma venenosum*
Bean, castor	*Ricinus communis*
Bean, lima	see Butter bean
Bean, locust	*Parkia filicoidea*
Chick pea	*Cicer arietinum*
Cow pea	*Vigna* spp.
Gram, black	*Phaseolus mungo*
Gram, green	*Phaseolus aureus*
Groundnut	*Arachis hypogea*
Lentil	*Lens culinaris*
Lupin	*Lupine* spp.
Pea	*Pisum sativum*
Soya	*Glycine max*
Tare	Lathyrus sativus

Fruits and vegetables

Almond	*Prunus dulcis*
Aubergine	see Egg plant
Avocado	*Persea americana*
Banana	*Musa*
Baobab	*Adansonia digitata*
Bitter leaf	*Vermonia amygdalina*
Brazil nut	*Bertholletia excelsa*
Breadfruit	*Artocarpus communis*
Cashew apple	*Anacardium occidentale*
Cinnamon	*Cinnamonum zeylanicum*
Coconut	*Cocos nucifera*
Coriander	*Coriandrum sativum*
Cottonseed	*Gossypium* spp.
Cumin	*Cuminum cyminum*

Date palm	*Phoenix dactylifera*
Drum stick	*Moringa oleifera*
Durian	*Durio zibethinus*
Egg plant	Solanum melongena
Garden egg	see Egg plant
Grapefruit	*Citrus paradisi*
Guava	*Psidium guajava*
Guinea sorrel	*Hibiscus sabdariffa*
Jack fruit	*Artocarpus integrifolia*
Kumquat	*Fortunella* spp.
Lemon	*Citrus limon*
Lime	*Citrus aurantifolia*
Mango	*Mangifera indica*
Niger seed	*Guizotia abyssinica*
Oil palm	*Elaeis guineensis*
Okra	*Hibiscus esculentus*
Olive	*Olea europaea*
Onion	*Allium* spp.
Orange	*Citrus sinensis*
Papaya (pawpaw)	*Carica papaya*
Peppers (sweet)	*Capsicum annum*
Pineapple	*Ananas comosus*
Plantain	*Musa paradisiaca*
Pumpkin	*Cucurbita* spp.
Rambutan	*Nephelium lappaceum*
Rapeseed	*Brassica napus*
Sago palm	*Metroxylon sagu*
Serendipity berry	*Dios-coreophyllum cumensii*
Sesame	*Sesamum indicum*
Star fruit	*Averrhoa carambola*
Sunflower	*Helianthus annuus*
Tangerine	*Citrus reticulata*
Tomato	*Lycopersicon esculentum*
Yeast (baker's)	*Saccharomyces cerevisiae*

Roots and tubers

Cassava	*Manihot utilissima*
Cocoyam	*Xanthosoma sagittifolium*
Potato, English	*Solanum tuberosum*
Potato, sweet	*Ipomoea batatas*
Taro	*Colocasia antiquorum*
Yam	*Dioscorea* spp.

Infusion drinks and sweeteners

Cocoa	*Theobroma cacao*
Coffee	*Coffea arabica*
Mate	*Ilex paraguensis*
Sugar, beet	*Beta vulgaris*
Sugar, cane	*Saccharum officinarum*
Sugar, maple	*Acer saccharum*
Sugar, palm	*Arenga saccharifera*
Tea	*Camellia sinensis*

Lower animals

Abalone	*Haliotis gigantea*
Crab	*Cancer spp.*
Crayfish	*Astacus* spp.
Cuttlefish	*Sepia* spp.
Lake fly	*Chaoborus*
Locust	*Nomadacris*
Sea cucumber	*Holothuria edulis*
Shrimp	*Pandalus* spp.
Snail, African	*Achatina* spp., *Archachatina* spp.
Squid	*Loligo* spp.

Fatty fish

Bonga	*Ethmalosa dorsalis*

Bream	*Tilapia mossambica*
Pacific mackerel	*Pneumatophorus japonicus*
Salmon	*Onchorhynchus garbuscha*
Sardine	*Sardinella longiceps*
Tuna	*Katsuwonus pelamis*

White fish

Bambangin	*Lutjanus fulvus*
Barracouda	*Shyraena jello*
Cod	*Gadus morhua*
Grouper	*Epinephalus undulosus*
Haddock	*Melanogramus aeglefinus*
Parang	*Chirocentrus dorab*
Puffer	*Sphaeroides spp.*

APPENDIX 2

Weights and measures

Length

1 mm	=	1000 μm	=	0.0394 inch
1 cm	=	10 mm	=	0.3937 inch
1 m	=	100 cm	=	1.0936 yards
1 km	=	1000 m	=	0.6214 miles
1 foot	=	12 inches	=	30.48 cm
1 yard	=	36 inches	=	0.9144 m
1 mile	=	1760 yards	=	1.6093 km

Area

1 sq m	=	10000 sq cm	=	1.196 sq yards
1 hectare	=	10000 sq m	=	2.4711 acres
1 sq km	=	100 hectares	=	0.3861 sq mile
1 sq foot	=	144 sq inches	=	0.0929 sq m
1 sq yard	=	9 sq feet	=	0.8361 sq m
1 acre	=	4840 sq yards	=	4046.9 sq m

Capacity

1 litre	=	1000 ml	=	0.22 gallons (UK)
1 pint	=	0.5683 litre		
1 gallon (UK)	=	8 pints (UK)	=	4.5461 litre
1 pint (US)	=	0.8327 pints (UK)	=	0.4732 litre
1 gallon (US)	=	0.8327 gallons (UK)	=	3.7853 litre
1 bushel (US)	=	0.9689 bushels (UK)	=	35.238 litre

Weight

1 g	=	1000 mg	=	0.0353 ounce
1 kg	=	1000 g	=	2.2046 pounds
1 tonne	=	1000 kg	=	0.9842 tons
1 pound	=	16 ounces	=	0.4536 kg

1 stone	=	14 pounds	=	6.35 kg
1 ton	=	2240 pounds	=	1.016 tonnes

Temperature conversion

$C = 5/9\ (F - 32)$ $\qquad$ $F = 9/5\ C + 32$

where C = degree Celsius and F = degree Fahrenheit.

Energy

1 kJ = 1000 joules
1 MJ = 1000 kJ
1 kilocalorie (kcal) = 4.184 kJ

Radiation

1 gray (Gy)	=	100 erg/g
1 kGy	=	1000 Gy
1 Gy	=	100 rad
1 krad	=	1000 rad
1 Mrad	=	1000 krad

APPENDIX 3

Rapid methods for the detection of aflatoxin

Aflatest

This is a rapid screening method for aflatoxins B_1, B_2, G_1 and G_2 in a wide range of commodities including groundnuts, palm kernel, maize, copra and cottonseed. The basis of this assay is an affinity column containing monoclonal antibodies which have the ability to isolate aflatoxin specifically from a crude sample of extract.

The sample is blended for 1 min in 60% methanol in water, then diluted with water to reduce the concentration of aflatoxin sufficiently to allow binding between the aflatoxin and the antibody. The extract is filtered and a 10-ml aliquot is applied to the top of the affinity column using a syringe provided.

The sample extract is slowly passed through the column where binding takes place between the aflatoxin and the antibody, retaining the aflatoxin in the affinity column.

The column is washed with distilled water to remove any extraneous unbound material from the column.

The bound aflatoxin is released from the antibody by passing pure methanol through the column.

The aflatoxin in the methanol eluted from the column can be determined in two ways:

(1) The eluate can be passed through a tip containing Florisil which binds the aflatoxin and enhances its natural fluorescence. The tips can be irradiated with UV light and compared to non-toxic fluorescent standards to provide a semiquantitative estimation of aflatoxin.

(2) Alternatively, a quantitative result can be obtained by reading the methanol eluate in a fluorimeter calibrated with non-toxic fluorescent solutions equivalent to 0, 10, 25 and 50 μg/kg aflatoxin.

Quantitox

The Quantitox is an Enzyme Immuno-assay (ELISA) based on a strip-plate format. The major advance in this system is the elimination of aflatoxin standard. A standard curve is provided, calibrated for each batch of kits. The assay utilises a simple methanol–water sample extraction and two 30-min incubations at room temperature and therefore provides a rapid, flexible method. The test is specific for aflatoxin B_1. These kits are available from May and Baker Diagnostics Ltd, 187 George Street, Glasgow, G1 1YT, UK.

FURTHER READING

Chapter 1

Ackroyd, W.R. *The Conquest of Famine*. Chatto & Windus, London, 1974.
Brittain, V. & Simmons, M. (ed.) *The Guardian Third World Review. Voices from the South*. Hodder and Stoughton, London, 1987.
FAO. *Commodity Review and Outlook 1985–86*. FAO Rome 1986.
UN Demographic Yearbook 1984. United Nations, New York, 1986.

Chapter 2

AOAC Official Methods of Analysis of the Association of Official Analytical Chemists. AOAC, Washington DC, 1980.
Berk, Z. (ed.) *Bravermans Introduction to the Biochemistry of Foods*. Elsevier, Amsterdam, 1976.
Gurr, M.I. *Role of Fats of Food and Nutrition*. Elsevier, Amsterdam, 1984.
Heimann, W. *Fundamentals of Food Chemistry*. Ellis Horwood, Chichester, 1980.
Paul, A.A. & Southgate, D.A.T. *The Composition of Foods*. HMSO, London, 1978.
Pearson, D. *Chemical Analysis of Food*. Churchill, Edinburgh, 1981.
Tan, S.P., Wenlock, R.W. & Buss, D.H. *Immigrant Foods*. HMSO, London, 1985.
WHO. *International Standards for Drinking Water*. WHO Geneva, 1971.

Chapter 3

Biological Sciences Curriculum Series, American Inst. of Biological Science, *Molecules to Man*. Edward Arnold, London, 1977.
Burke, S.R. *Human Anatomy and Physiology for the Health Sciences*. Wiley, Chichester, 1980.
Simpson, G.G. & Beck, W.S. *Life: An Introduction to Biology*. Harcourt Brace Jovanovich Ltd, London, 1970.

Chapter 4

Anazonwu-Bello, J.N. *Food and Nutrition in Practice*. Macmillan, London, 1976.
Burton, B.T. *The Heinz Handbook of Nutrition*. McGraw-Hill, London, 1976.
Latham, M.C. *Human Nutrition on Tropical Africa*. FAO, Rome, 1979.
Ministry of Agriculture, Fisheries and Food. *Manual of Nutrition*. HMSO, London, 1979.
Passmore, R. & Eastwood, M.A. *Human Nutrition and Dietetics*. Churchill Livingstone, Edinburgh, 1986.
Platt, B.S. *Tables of Representative Values of Foods Commonly used in Tropical Countries*. HMSO, London, 1976.

Chapter 5

Amerine, M.A., Pangborn, R.M. & Rossler, E.B. *Principles of the Sensory Evaluation of Food*. Academic Press, London, 1965.
Chan, H.W. (ed.) *Biophysical Methods in Food Research*. Blackwell, Oxford, 1984.
Francis, F.J. & Clydesdale, F.M. *Food Colorimetry: Theory and Applications*. AVI, Westport Connecticut, 1975.
Jellinek, G. *Sensory Evaluation of Food*. Ellis Horwood, Chichester, 1985.
McKinney, G. & Little, A.G. *Colour of Foods*. AVI, Westport, Connecticut, 1962.
Sherman, P. *Food Texture and Rheology*. Academic Press, London, 1979.
Vaughan, J.G. (ed.) *Food Microscopy*. Academic Press, London, 1979.

Chapter 6

Frazier, W.C. *Food Microbiology*. McGraw-Hill, New York, 1978.
Hayes, P.R. *Food Microbiology & Hygiene*. Elsevier, London, 1985.
Parry, T.J. & Pawsey, R.K. *Principles of Microbiology for Students of Food Technology*. Hutchinson, London, 1973.
Stanier, R.Y., Adelberg, E.A. & Ingraham, J.L. *General Microbiology*. Macmillan, Basingstoke, 1976.

Chapter 7

Ackroyd, W.R. & Doughty, J. *Legumes in Human Nutrition*. Academic Press, London, 1978.
Houston, D.F. *Rice*. American Association of Cereal Chemists, St Pauls, Minnesota, 1972.
Hulse, J.H., Laing, E.M. & Pearson, D.E. *Sorghum and the Millets: Their Composition and Nutritive Value*. Academic Press, London, 1980.
Inglett, G.E. *Corn: Culture, Processing, Products*. AVI, Westport, Connecticut, 1970.
Inglett, G.E. & Munck, C.K. *Cereals for Food and Beverages*. Academic Press, London, 1980.
Kent, N.L. *Technology of Cereals*. Pergamon, Oxford, 1975.
Lockwood, J.F. *Flour Milling*. Northern Publishing Co. Ltd, London, 1962.

Muller, H.G. *Baking and Bakeries*. Shire Publications, Princes Risborough, 1986.
Williams, A. *Breadmaking, the Modern Revolution*. Hutchinson, London, 1975.

Chapter 8

Brouk, B. *Plants Consumed by Man*. Academic Press, London, 1975.
Coursey, D.G. *Yams*. Longmans, Harlow, 1967.
Doku, E.V. *Cassava in Ghana*. Ghana University Press, Legon, 1969 (also Panther edn, 1971).
Duckworth, R.B. *Fruit and Vegetables*. Pergamon, Oxford, 1966.
Hartley, C.W.S. *The Oil Palm*. Longmans, London, 1967.
Nagy, S. & Shaw, P.E. *Tropical and Subtropical Fruits: Composition, Properties and Uses*. AVI, Westport, Connecticut, 1980.
Onunema, I.C. *The Tropical Tuber Crops*. Wiley, Chichester, 1978.
Pantastico, E.R.B. *Post-harvest Physiology, Handling and Utilisation of Tropical and Subtropical Fruits and Vegetables*. AVI, Westport, Connecticut, 1975.
Reed, G. *Enzymes in Food Processing*. Food Science & Technology Series, Academic Press, New York, 1975.
Vickery, M.L. & Vickery, B. *Plant Products of Tropical Africa*. Macmillan, London, 1979.

Chapter 9

Eden, T. *Tea*. Longmans, London, 1965.
Minifie, B.M. *Chocolate, Cocoa and Confectionery: Science and Technology*. AVI, Westport, Connecticut, 1980.
Nelson, P.E. & Tressler, D.K. *Fruit and Vegetable Juice Processing Technology*. AVI, Westport, Connecticut, 1980.
Sivetz, M. & Elliot Foote, H. *Coffee Processing Technology*. AVI, Westport, Connecticut, 1963.
Thorner, M.E. & Herzberg, R.J. *Food Beverage Service Handbook*. AVI, Westport, Connecticut, 1970.
Ward, G.A.R. *Cocoa*. Longmans, London, 1975.

Chapter 10

Birch, G.G., Green, L.F. & Coulson, C.B. *Sweetness and Sweeteners*. Applied Science Publishers, Barking, Essex, 1971.
Birch, G.G. & Parker, K.J. *Sugar: Science and Technology*. Applied Science Publishers, Barking, Essex, 1979.

Chapter 11

Henrickson, R.L. *Meat, Poultry and Seafood Technology*. Prentice-Hall, Hemel Hempstead, 1978.
Lawrie, R.A. *Meat Science*. Pergamon, Oxford, 1985.
Love, R.M. *Chemical Biology of Fishes*. Academic Press, London, 1970.

Mann, I. *Meat Handling in Under-developed Countries. Slaughter and Preservation*. Agricultural Development Paper No. 70, FAO, Rome, 1960.
Robinson, R.K. *Modern Dairy Technology*, vols. 1, 2. Elsevier, Amsterdam, 1986.
Stadelman, W.J. & Cotterill, O.J. *Egg Science and Technology*. AVI, Westport, Connecticut, 1973.
Wilson, N.R.P. *Meat and Meat Products; Factors Affecting Quality Control*. Applied Science Publishers, Barking, Essex, 1981.

Chapter 12

Factories Act 1961. HMSO, London, 1975.
O'Keefe, J.A. *Bell and O'Keefe's Sale of Food and Drugs*. Butterworth, Sevenoaks, Kent, 1968.

Chapter 13

Brennan, J.G., Butters, J.R., Cowell, N.D. & Lilly, A.E.V. *Food Engineering Operations*. Applied Science Publishers, Barking, Essex, 1976.
Coulson, J.M. & Richardson, J.F. *Chemical Engineering*. Pergamon, Oxford, 1977, vol. 1; 1978, vol. 2.
Earle, R.L. *Unit Operations in Food Processing*. Pergamon, Oxford, 1983.
Harris, R.S. & Karmas, E. *Nutritional Evaluation of Food Processing*. AVI, Westport, Connecticut, 1975.
Høyem, T. & Kvåle, O. *Physical, Chemical and Biological Changes in Food caused by Thermal Processing*. Applied Science Publishers, Barking, Essex, 1977.
Jenkins, G.H. *Introduction to Cane Sugar Technology*. Elsevier, Amsterdam, 1966.
Pirie, N.W. (ed.) *Leaf Protein: Its Agronomy, Preparation, Quality and Use*. Blackwell, Oxford, 1971.

Chapter 14

Hesseltine, C.W. & Wang, H.L. *Indigenous Fermented Food of non-Western Origin*. Mycologia Memoir No. 11. Cramer, Berlin, 1986.
Hough, J.S., Briggs, D.E. & Stevens, R. *Malting and Brewing Science*. Chapman & Hall, London, 1971.
Steinkraus, H.K. *Handbook of Indigenous Fermented Foods*. Marcel Dekker, New York, 1983.

Chapter 15

Desrosier, N.W. & Desrosier, I.N. *Technology of Food Preservation*. AVI, Westport, Connecticut, 1977.
Hersom, A.C. & Hulland, E.D. *Canned Foods; Thermal Processing and Microbiology*. Churchill-Livingstone, Edinburgh, 1980.
Irradiation of Food: Six Papers. IFST (London). *Proceedings*, vol. **19**, No. 4, December 1986.

Josephson, E.S. & Peterson, M.S. *Preservation of Food by Ionising Radiation*, vol. 1–3. CRC Press, Boca Raton, 1982–3.
Jul, M. *The Quality of Frozen Food*. Academic Press, Westport, Connecticut, 1984.
Report on the Safety and Wholesomeness of Irradiated Foods. Advisory Committee on irradiated and novel foods. HMSO, London, 1986.
Thorne, S. *Developments in Food Preservation*, 2 vols. Applied Science Publishers, London, 1981–3.
Tressler, D.K., van Ardsel, W.B. & Copley, J.J. *The Freezing Preservation of Foods*, vols. 1, 2, 3, 4. AVI, Westport, Connecticut, 1968.
Yaciuk, G. *Food Drying*. IDRC OHawa, 1981.

Chapter 16

Reed, G. *Enzymes in Food Processing*. Academic Press, London, 1975.
Taylor, R.J. *Food Additives*. Wiley, Chichester, 1980.
Whitaker, J.R. *Principles of Enzymology for the Food Sciences*. Marcel Dekker, New York, 1972.

Chapter 17

Alleyne, G.A.O., Hay, R.W., Picon, D.I., Stanfield, J.P. & Whitehead, R.G. *Protein–Energy Malnutrition*. Arnold, London, 1977.
Howe, G.M. (ed.) *A World Geography of Human Diseases*. Academic Press, London, 1977.
McLaren, D.S. *Nutrition and Its Disorders*. Churchill Livingstone, Edinburgh, 1976.

INDEX